ELECTRICAL ESTIMATING

PROFESSIONAL REFERENCE

Paul Rosenberg

Created exclusively
for DEWALT by:

www.palpublications.com
1-800-246-2175

Titles Available From DeWALT

DeWALT Trade Reference Series

Construction Professional Reference

Datacom Professional Reference

Electric Motor Professional Reference

Electrical Estimating Professional Reference

Electrical Professional Reference

HVAC Professional Reference

Lighting & Maintenance Professional Reference

Plumbing Professional Reference

Referencia profesional sobre la industria eléctrica

Security, Sound & Video Professional Reference

Wiring Diagrams Professional Reference

DeWALT Exam and Certification Series

Electrical Licensing Exam Guide

HVAC Technician Certification Exam Guide

 Pal Publications, Inc., a licensee of DeWALT Industrial To[illegible] 374 Circle of Progress Drive, Pottstown, PA 19464, Tel.: 800-246-217[illegible]

This Book Belongs To:

Name: ______________________________

Company: ______________________________

Title: ______________________________

Department: ______________________________

Company Address: ______________________________

____

____

Company Phone: ______________________________

Home Phone: ______________________________

Pal Publications, Inc.
374 Circle of Progress
Pottstown, PA 19464-3810

First edition published 2005

ISBN 0-9770003-0-3

09 08 07 06 05 5 4 3 2 1

Printed in the United States of America

A Note To Our Customers

We have manufactured this book to the highest quality standards possible. The cover is made of a flexible, durable and water-resistant material able to withstand the toughest on-the-job conditions. We also utilize the Otabind process which allows this book to lay flatter than traditional paperback books that tend to snap shut while in use.

Electrical Estimating Professional Reference is not a substitute for the National Electric Code®. National Electric Code® and NEC® are registered trademarks of the National Fire Protection Association, Inc. Quincy, MA.

Preface

Electrical estimating is a critical skill for anyone who charges for an electrical job. Knowing how to install wiring is a nice skill to have, but charging for the use of this skill requires a knowledge of estimating. Unfortunately, very few of us are taught this skill in apprenticeship school.

For a long time there has been a need for a short book in electrical estimating that explains the process in clear, plain language. Our primary goal in putting this book together was that it should be written so that any electrician could pick it up, understand it, and use it. Actually, that is how all electrical books should be written, but for some reason estimating books tend to be especially difficult.

Chapter One of this book covers itemized estimating. This is the standard method of electrical estimating, and the only one that works for every type of job. If you had to understand only one type of electrical estimating, this would be the one. Itemized estimating is the basic method of estimating and all other forms require you to understand this first.

Chapter Two covers unit pricing, which is a good estimating technique for residential and small commercial projects. Unit pricing can save you time on these types of estimates, if you use it correctly.

Chapters Three and Four cover communication and electronic installations. These projects are unique in that they require an unusual amount of interconnection. A programmable controller, for example, may be fairly easy to install by itself, but interconnecting it with the devices it controls could take anywhere from one hour to several days. We have devoted a section of this book to the most common types of high-tech installations.

Finally, Chapter Five covers algebraic estimating. This will be of use primarily to full-time estimators. But, those who learn the process will be able to estimate at unheard of speeds.

Electrical estimating takes time and effort, but the basic process is certainly not too much for a competent electrician.

Best wishes,
Paul Rosenberg

CONTENTS

Installation Labor Units *(cont.)*

Installation Labor Units *(cont.)*

Installation Labor Units *(cont.)*

Installation Labor Units *(cont.)*

Installation Labor Units ***(cont.)***

CHAPTER 2 – *Unit Pricing* 2-1

CHAPTER 5 – *Algebraic Estimating* . . . 5-1

CHAPTER 1
Itemized Estimating

The basic purpose of electrical estimating is to take a set of documents (a set of construction drawings and a book of specifications) and translate the information in the documents into a total dollar price for the work. This process is usually divided into three basic parts:

1. The takeoff.

2. Writing up the estimate.

3. Summarizing the estimate.

The process of "taking off" the job is literally "taking" the information "off" the plans and transferring it to separate sheets. The estimator interprets the graphic symbols on the plans and translates them into words and numbers, which can then be processed.

In "writing up" the estimate, the estimator transfers the takeoff information to special sheets, assigns both material and labor prices to each item and totals these prices on each sheet.

In the summary, the estimator adds all the pricing sheets to give a total cost for the material and labor for the project being estimated. Any other costs that will be required for the project's completion, overhead for the maintenance of the company's internal operations and profit must also be included to determine a final selling price.

The basic purpose of estimating is to get from the plans to the price.

The itemized estimate is the standard of the industry and has been for many years. This chapter provides a complete explanation of the specific steps that the estimator will have to follow to be proficient at this technique. Itemized estimating requires the use of **labor units**. At the end of this chapter are labor units for the most commonly used electrical materials. These units represent the amount of time that should be required to install each item during a normal installation. These units are shown as hours per unit or hours per foot, as the case may be.

For example, ½ in. EMT conduit, under normal circumstances, will be installed at the rate of .02 hours per foot. Note that this is the labor required for installing the conduit only and that the required connectors, couplings and straps are labored separately. You see also that No. 12 wire can be expected to be installed at a rate of .005 hours per foot.

It is also important to note that the estimator must use his or her own good judgement, developed through experience, to determine when a specific installation is significantly more difficult than the average. In such instances, the labor units for that particular item or items must be increased. Usually, an increase of 20 to 30 percent is enough, but some very difficult circumstances require even higher labor units.

One final point must be made: the quality of any type of electrical estimating is completely dependent on the skill, diligence and judgement of the estimator.

This book was designed to give the reader skills and knowledge required to prepare good estimates. To this, the estimator must add diligence and good judgement. And the more experience the better.

BASICS

Itemized estimating is the process of itemizing and pricing every piece of material that goes into an electrical construction installation in order to determine the cost of that installation. An itemized estimate is the surest way to get a complete and accurate list of materials that will be required for any particular installation. There is no type of estimate quite as reliable as a well prepared itemized estimate.

In order to do an itemized estimate thoroughly and efficiently, a careful procedure must be followed by the estimator. First the estimator should have a quiet and well lit area to work in, free from distractions as much as possible. The estimator must then follow a step-by-step procedure to assure that he or she has taken all the required information off the plans and transferred this information to a pricing sheet where it can be accounted for.

STEPS IN ITEMIZED ESTIMATING

The recommended procedure for preparing an itemized estimate is as follows:

1. Review the job you are about to estimate. Make a thorough review of the plans and specifications for the installation. Get a good understanding of exactly what the building (or whatever) is going to be like. Review the entire set of plans and specs (specifications), not just the electrical pages. The first section of the specs, usually called "General Conditions," spells out exactly how the job will be run, how payments will be made, how disputes will be settled, and so on. This first section of the specifications is especially important. Section 15 of the specs, which governs the installation of the mechanical equipment, is also important, because the electrical and mechanical trades have a great deal of overlap in their work. Go over Section 16 completely and thoroughly, as it is the section that explains the specific techniques and materials required for the electrical installation.

Starting on **page 1-12, Figure 1.1** shows a prebid checklist that you should fill out as you review the plans and specifications. The items shown on this checklist are the most important considerations to the estimator in pricing the installation.

Take special care in reviewing the site drawings which may contain expensive details of the electrical installation that are not specifically shown elsewhere. Often you will find site lighting, service runs and remote devices on these sheets. Review all the plan sheets, but give special attention to the sheets that show work with which your work must interface, most typically mechanical and ceiling work.

Take a good look at **Figure 1.2**, which starts on **page 1-14**. It is a list of very important questions that you will want to think about as you prepare your estimate. You should also go over these questions as you price and review your estimate as well. Use this list as a checklist to make sure that you have covered all the expenses that you will incur during the job you are estimating. Almost all the mistakes made in itemized takeoffs are omissions, so a careful review of the plans and specs and repeated reference to a good checklist are extremely important.

2. Do a takeoff of the lighting fixtures. Starting on the first electrical sheet, completely color in and count every lighting fixture. Do one type of fixture at a time and use a hand-held counter. Check the fixture schedule on the plans to make sure that you understand and count every type of fixture. Write down all the totals on a separate sheet of paper. Leave plenty of blank space on this tabulation sheet. During the following stages of your takeoff, you are bound to find stray light fixtures that you missed the first time through, and you will need room to record these additional fixtures.

3. Do a takeoff of all the wiring devices. With your hand-held counter in one hand, color in all the single-pole switches on the drawings. As you count the single-pole switches, also count the other types of switches by using hash marks on the side of the plans or on a separate sheet of paper. While you do this, also keep track of how many two-gang rings will be required by using hash marks for them. On **page 1-19, Figure 1.3** shows how this is done. This figure shows a portion of an office building. There are different types of switches scattered throughout, which all need to be counted in your takeoff. For this drawing, the estimator, with counter in hand, has counted (and colored in) all the single-pole switches. At the same time, he colored in any of the three-way or dimmer switches he found and made hash marks for them on the side of the plans. Whenever he ran across two or three switches together in the same box, he noted them with hash marks also, so that the takeoff includes the right types of boxes and plaster rings. You see in **Figure 1.3** that the estimator found a total of 10 single-pole switches (the first of the five quantities recorded at the bottom of the drawing). To the right, notice that the estimator used hash marks to record 5 three-way switches as he went along. To the right of this, he used hash marks to record 2 dimmer switches. Again to the right, he recorded that *of these switches that he counted*, there were 4 pairs in a two-gang configuration. On the far right, the estimator recorded that there was one set of three switches grouped together.

At the bottom of **Figure 1.3** is a tabulation of the types of plaster rings that will be needed for the switches shown in this drawing. The quantities of two- and three-gang rings are easily obtained by just looking at the hash marks below the drawing. In order to get the proper number of single-gang rings, we must first find the total number of switches of all types. This total comes to 17 (10 single-pole, 5 three-way, 2 dimmer). From this total of 17, we find that 3 are located in a three-gang enclosure and that 8 (4 sets of 2) are located in two-gang enclosures. By subtracting 11 (3 + 8) from the total of 17, we find that we will need 6 one-gang rings. This simple mathematical process is called "determining a quantity from a quantity."

4. Do a takeoff of the one-line diagram. The one-line diagram shows the power distribution for the project you are estimating. Carefully make a list of each item of equipment that is shown on this diagram, and write a good description of each item. Next, you must do a good takeoff of the various feeder circuits that bring power to these pieces of equipment. The feeder schedule shown in **Figure 1.4**, which can be found on **page 1-20**, is ideal for this use. Each feeder shown on the one-line diagram should be recorded on this schedule; and as it is recorded, it should be completely colored. All the equipment shown on the one-line diagram should also be completely colored in. As you take off all these feeders, you will have to find the various items on the plans in order to determine the distances between them. Measure the distances between the items with a rotometer, also sometimes called a map reader. Make sure that you are using the right scale for each drawing, and be sure to include the vertical portions of the feeder runs, which cannot be shown on two-dimensional drawings.

5. Do a takeoff of all the branch circuitry, including raceways and wire. This is the most lengthy part of the takeoff procedure, although it is not exceptionally difficult. Starting with the smaller sizes of raceway, color in each raceway and wire combination with a separate color. For instance: ½ in. EMT with two No. 12 THHN wires could be light blue, ½ in. EMT with three No. 12 wires could be bright red, ½ in. EMT with four No. 12 wires could be brown, ¾ in. EMT with two No. 12 wires could be purple, and so on. By the time you run out of colors (assuming you have a good set of colored pencils), you should have indicated all the common types of raceway and wire combinations and can simply note what the remaining raceways are.

As you color these runs, keep a counter in your other hand, and count the connectors that will be required (one on each end of every run). Note this quantity on a separate sheet of paper. Next, with your rotometer, measure the length of all the runs of each type of raceway and wire combination. As you measure these runs, keep your counter in your other hand and click it every time the run you are measuring goes vertical, either up or down. When you finish all the runs of one color, write down the total measurement and add to it the vertical distances as well. This will give you the proper total for this combination of raceway and wire. The correct vertical distances for each project can be calculated by looking at the wall cross sections on the architectural plans and the outlet mounting heights shown on the electrical plans.

After you finish taking off every type of raceway and wire combination shown on the plans, you should have a sheet that looks something like the one in **Figure 1.5**, which can be found on **page 1-21**. All the quantities taken off the plans should be written on a sheet like this and then totaled, as is shown at the bottom of the sheet. When these totals are transferred to an estimate sheet for the final pricing, it is usually best to increase the quantities of the conduit slightly (2 to 4 percent) to account for waste and to increase the quantities of wire about 5 percent to account for the wire that must be left hanging out of boxes or that is wasted.

6. Do a takeoff of all the special systems. Most medium- to large-sized electrical installations will require one or more types of special electrical or electronic systems. The estimate will be required to include the cost of these systems. These systems include any unusual type of system, such as fire alarms, lightning protection, sound systems, closed circuit TV, nurse call systems, and so on.

Special systems are usually handled in one of two ways:

a. The estimator does a complete takeoff of all the individual items required by the special system, prices each item, and when the time comes, the company electricians will install the system.

b. The estimator reviews the special system(s) shown on the plans and then calls specialty contractors to get prices for the system(s) in question. The estimator reviews these prices, compares them with the plans and incorporates them into the final estimate. At the appropriate time, the specialty contractor will send workers to install the system and will be responsible for its correct installation and acceptance by the owner.

7. Write up the estimate. Your next job is to write on a pricing sheet all the quantities you have been keeping on separate sheets of paper. A typical pricing sheet is shown in **Figure 1.6**, which starts on **page 1-22**. On this sheet, you first write down the quantities of all the electrical materials that will be required for the installation. Then you fill in the appropriate material and labor prices for each item and extend these figures. A completed pricing sheet is shown in **Figure 1.7**, which can be found on **page 1-24**. When writing up EMT conduit, you can use the "quantity from a quantity" method to obtain the proper number of couplings (one coupling for every 10 feet of conduit) and straps (one strap for every 10 feet of conduit, rounded up a little bit).

8. Finalize the estimate. This is the final step in preparing the itemized estimate. Review the plans and the estimate and pricing sheets, making sure that everything has been included. Next, take quotes from suppliers on materials that you cannot price. Most typically, these items are light fixtures, switchgear and special systems. After you receive and review these quotes, record them on a bid summary sheet as shown in **Figure 1.8**, which can be found on **page 1-25**. Record on the bid summary sheet on the line marked "miscellaneous materials" the total material price taken from the pricing sheets. Also record on the bid summary sheet the labor hour figures taken from the pricing sheets. Multiply these by the average hourly labor rate to obtain a total cost of labor for the installation. Add all these items together with job costs, overhead and profit to get a final bid price. Job costs typically include the following:

- **Storage trailer rental and insurance.**
- **Equipment rental.**
- **Job site utilities.**
- **Blueprinting costs.**
- **Travel and lodging expenses.**
- **Performance bonds.**
- **Any other costs that will be incurred because of this project, but not included in the company's overhead.**

FIGURE 1.1 — PREBID INFORMATION

Bidding Information

Name of Job ____________________

Location ____________________

Bid Date ____________________

Bids to ____________________

Design Information

Architect ____________________

Engineer ____________________

Owner ____________________

Financial Information

Bid Bond ____________________

Construction Time ____________________

Liquidated Damages ____________________

Retainage ____________________

Wage Rate ____________________

Construction Information

Building ____________________

Floors ____________________

Walls ____________________

Ceilings ____________________

Elevations ____________________

Quotes Required

Lighting Fixtures ____________________

Distribution Apparatus ____________________

Fire Alarm ____________________

Clock ____________________

TV ____________________

Sound ____________________

Intercom ____________________

Nurse Call ____________________

Doctor's Register ____________________

Security ____________________

Specifications

Devices

Switches ____________________

Receptacles ____________________

Wall Plates ____________________

Special Outlets ____________________

Branch Wire

Type ____________________

Separate Ground ____________________

Lightning Protection ____________
Electric Heat ____________
Snow Melting ____________
Heat Tracing ____________
Emerg. Generators ____________
Dimming ____________
Stage Lighting ____________
PCs ____________
Energy Management ____________
Signal ____________
Isolating Panels ____________
Cable Tray ____________
Flexible Wiring ____________
U.F. Duct ____________

Feeder Wire
Type ____________
Separate Ground ____________
High Voltage ____________

Raceways
Underground ____________
Slab ____________
Surface ____________
Partitions ____________
Fittings ____________
Minimum Size ____________
Supports ____________
Boxes ____________

Cost Responsibility

Temporary Electrical ____________
Excavation & Backfill ____________
Concrete Encasement ____________
Equipment Pads ____________
Pole Bases ____________
Fixture Support ____________
Cutting & Patching ____________
Pitchpockets ____________

Sleeves ____________
Painting ____________
Starters ____________
Disconnects ____________
Control Wiring ____________
Interlock Wiring ____________
Primary Transformer ____________
Primary Raceway ____________
Primary Cable ____________

FIGURE 1.2 — QUESTION LIST FOR PREPARING ESTIMATE

Who is the owner?

Is there anything about the owner that will make the job easier or harder than usual?

Who is the architect?

Does the architect have any particular characteristics that you should consider?

Who is the engineer?

Does the engineer have any particular characteristics that you should consider?

Will you be expected to furnish temporary power to the job site?

Who pays the electric bill?

Will you be required to bring telephone service to the site?

Will you need to bring power and/or phone service to any construction trailers?

When is the bid date?

Are there any special requirements for submitting your bid?

How many alternate bids are there?

How many of these alternates affect you?

How will payments be made?

When will payments be made?

FIGURE 1.2 — QUESTION LIST FOR PREPARING ESTIMATE *(cont.)*

How much retainage will be withheld?

Will the retainage be reduced during the course of the project?

How will change orders be handled?

Will you be held to any certain percentage of markup on change orders?

Who decides what necessitates a change order and what doesn't?

How much time is being given to complete the project?

Will there be liquidated damages if the job goes past the completion date?

Who decides who pays the liquidated damages?

How will any disputes be handled?

Will you need to figure any overtime work into the job to get it completed in time?

Will you be allowed to substitute any types of materials for others?

Will you be required to post a bond?

Will you be required to carry certain types or certain amounts of insurance?

Do you need to include an allowance for a bond or extra insurance?

FIGURE 1.2 — QUESTION LIST FOR PREPARING ESTIMATE *(cont.)*

Are there any special wage rates or labor requirements that you should allow for?

Who will be expected to supply and install disconnect switches on the mechanical equipment?

Who will be expected to supply starters and controllers for the mechanical equipment?

Who will be expected to mount and/or wire these starters and controllers?

Who is required to do the control wiring for the mechanical systems?

Are there any special forms or reports that you will be required to file?

Can you get paid for any material stored on the job site?

Are there any other special or unusual requirements?

How much time will you have to spend giving maintenance instructions to the owner's people?

What kind of as-built drawings will you be expected to furnish?

Will you be allowed to do any redesign? (Combining branch circuits, etc.)

Are there any special requirements for the switchgear?

FIGURE 1.2 — QUESTION LIST FOR PREPARING ESTIMATE *(cont.)*

Are there any special requirements for the control equipment?

What types and brands of light fixtures are specified?

Are there any alternate brands given?

Will you be allowed to make any substitutions?

Is there any special testing that you will be required to do?

Are there any items that you need to clarify with the engineer before you bid the job?

What types of wiring devices will be required?

Are there any special materials or methods of installation that you must use?

What special systems (if any) will be required?

What types of wiring systems will be used? (Rigid conduit, EMT, aluminum conduit, ENT, cable, etc.)

What codes will govern the installation?

Who will be inspecting your work?

How will these inspections be handled?

What types or brands of material will need to be used?

What types of fittings will be required?

FIGURE 1.2 — QUESTION LIST FOR PREPARING ESTIMATE *(cont.)*

What types of materials must you provide submittals for?

How many copies of submittals will you need to furnish?

How long a warranty period is specified?

Will you be required to do any maintenance?

Will you be required to supply any extra equipment? (Extra sets of fuses, etc.)

Will any special types of equipment be required? (Crane, backhoe, trencher, etc.)

Will there be any problems with congestion at the job site?

Will there be any problems storing or moving materials on the job site?

FIGURE 1.3 — TAKEOFF OF WIRING DEVICES

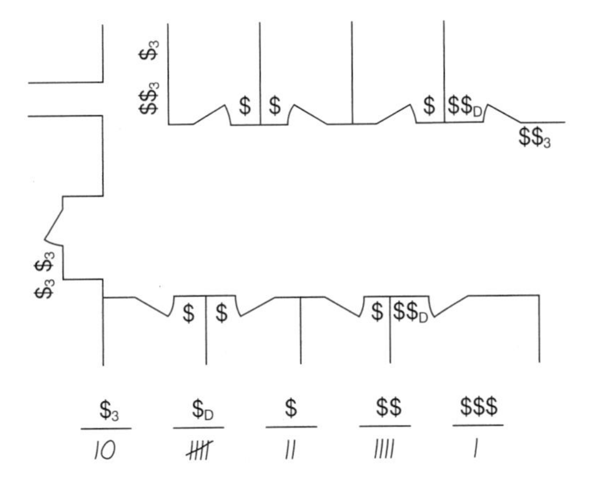

1 gg rings — 6
2 gg rings — 4
3 gg boxes & rings — 1

FIGURE 1.4 — FEEDER SCHEDULE

Job Name: ____________________ Date: __________

ROUTING		RACEWAY									WIRE				MISC.		
FROM	TO	SIZE	NO.	LENGTH	FOOTAGE	TYPE	90°ELL	TERMS	STRAPS	TRENCH	NO.	SIZE	LENGTH	FOOTAGE			

FIGURE 1.5 — WIRE AND RACEWAY TAKEOFF

Raceway/Wire	Connector Totals
4,190' ½" EMT – 2 #12	984
2,740' ½" EMT – 3 #12	
3,160' ½" EMT – 4 #12	
910' ¾" EMT – 3 #12	210
720' ¾" EMT – 4 #10	
480' ¾" EMT – 2 #10	
220' ¾" EMT – 3 #10	
1,980' 1" EMT – 3 #8	24
390' 1¼" EMT – 3 #3	10

Raceway/Wire Totals

Raceway		Wire	
½" EMT –	10,090'	#12 THHN –	31,970'
¾" EMT –	2,330'	#10 THHN –	4,500'
1" EMT –	1,980'	#8 THHN –	5,940'
1¼" EMT –	390'	#3 THHN –	1,170'

FIGURE 1.6 — PRICING SHEET

PRICING SHEET

Job ______________________________ Page ________

______________________________ of ________ Pages

____

____

Estimated by ______________ Checked by ______________ Date ________

Description	Material				Labor		
	Quantity	Unit Price	Per	Amount	Unit	Per	Amount

Totals

FIGURE 1.7 — A COMPLETED PRICING SHEET

PRICING SHEET

Job James Bay Warehouse **Page** 2 **of** 4 **Pages**

Estimated by MR **Checked by** PR **Date** 3-31-86

Description	Material				Labor		
	Quantity	Unit Price	Per	Amount	Unit	Per	Amount
1/2 EMT	1,200	13 50	C	162 —	.02	E	24 —
connector	174	9 80		17 05	.06		10 44
cplg.	120	10 08		12 10	.03		3 60
one hole strap	140	5 20		7 80	.03		4 20
3/4	150	20 90		31 35	.03		4 50
conn.	4	17 40		70	.06		2 40
cplg.	15	19 —		2 85	.03		45
one hole strap	17	7 70		1 31	.03		51
#14 THHN stranded	500	27 90	M	13 95	.004		2 —
#12 solid	3,000	31 —		93 —	.005		15 —
stranded	1,000	38 20		38 20	.005		5 —
			Totals	380 21			72 10

FIGURE 1.8 — BID SUMMARY SHEET

BID SUMMARY

Project:	Bid Date:
Misc. Material	$
Fixtures	$
Gear	$

Material Cost $ ________

Sales Tax $ ________

Total Materials $ ________

Total labor cost:

Hours @ $ ______ = $ ________

Job Costs (________) $ ________

Net Cost $________

% O & P ________

Total $________

Bid @ $________

INSTALLATION LABOR UNITS

Applying labor units to the quantities of electrical materials shown on the pricing sheets is extremely important to the process of electrical estimating. The estimator must start with a good set of labor units and apply them according to the conditions of the job he or she is estimating. This requires some tempering of the labor units. For instance, if a certain project has an especially high ceiling, the estimator will have to raise the labor units for any equipment that must be installed on or above the ceiling. The tempering of the labor units is what makes an estimate accurate. Careful thought and experience are required to temper labor units well.

At the end of this chapter you will find a well developed set of labor units. These units obviously cannot cover every job, company or job site condition, but they are a good starting point from which the estimator must take over. All the labor units are shown as hours per each, hours per foot, etc.

Notes to Labor Units

1. All pieces of equipment weighing over 250 pounds should be individually considered in assigning a labor unit. The estimator should mentally review the difficulty of putting the piece of equipment in place.

2. When groups of feeders are run in parallel, the labor units for installing both conduit and wire can be reduced by up to 30 percent, because of the reduced setup time for every run after the first.

3. Labor units for feeder wire should be increased if the runs are exceptionally short or decreased if the runs are exceptionally long. This is because of the ratio of setup time to the number of feet in the run. A short feeder run requires almost the same amount of setup time as does a long run.

4. The estimator should remember that it is sometimes prudent to apply overhead and profit separately to very expensive items (most commonly generators or subcontract work), as these items carry almost no risk, being primarily items simply bought and resold by the contractor. The percentages of overhead and profit can be lowered on these items.

ASSEMBLY ESTIMATING

The assembly method is based upon the concept that virtually every type of electrical symbol used in construction drawings can be summarized as a specific list of materials.

For example, the assembly for a common receptacle would include the receptacle, the finish plate, a box, a plaster ring, screws for fastening the box to the framing, a grounding pigtail, a couple of feet of No. 12 wire in the box, an average of two wire nuts, and two ½ in. EMT connectors. Thus the assembly includes everything indicated by the duplex receptacle symbol on the plans.

This is typical of all assemblies of which there could be many thousands of combinations of different types of receptacles with different types of finish plates, different types of plaster rings, and so on. Even raceways can be broken down into assemblies. For example, three No. 12 THHNs in ½ in. EMT, three No. 6 and one No. 10 in 1 in. EMT, and so on.

When the estimator takes off a job, he or she counts all the symbols and raceways on the plans and then prices the list as assemblies (so many type XYZ assemblies, so many type ABC assemblies, etc.). Since all the assemblies are prepriced and prelabored, the counting and pricing of individual parts are mostly eliminated.

While the assembly concept can be a great time-saver for electrical estimating, the process of developing and maintaining a full set of assemblies can be almost a full-time job. Several services offer assembly data for estimating. These services save a great deal of time in developing all the assemblies and in tabulating the material and labor costs. Care must be taken, however, with the material cost figures that are supplied by the services. Because of the services' broad spectrum of clients, they cannot publish material prices that are correct for everyone. They use the published contractor prices, which are considerably higher than what most contractors will actually pay. Therefore, an estimator who uses the material prices supplied by the services must take into account that the estimate will be too high.

If you look through the unit pricing estimate sheets in **Chapter 2**, you will see that the units shown are essentially assemblies with the associated raceway and wire added.

INSTALLATION LABOR UNITS

GRC Conduit

Size	Hours per Foot
½ in.	.03
¾	.04
1	.05
1¼	.06
1½	.07
2	.09
2½	.11
3	.14
3½	.17
4	.20
5	.25
6	.33

GRC Elbows — 45° and 90°

Size	Hours per Elbow
½ in.	.20
¾	.25
1	.30
1¼	.38
1½	.40
2	.45
2½	.62
3	.85
3½	.98
4	1.20
5	1.60
6	2.00

GRC Couplings

Size	Hours per Coupling
½ in.	.15
¾	.19
1	.24
1¼	.27
1½	.29
2	.32
2½	.50
3	.60
3½	.70
4	.80
5	.90
6	1.00

IMC Conduit

Size	Hours per Foot
½ in.	.02
¾	.03
1	.04
1¼	.05
1½	.06
2	.07
2½	.11
3	.13
3½	.15
4	.17

IMC Elbows — 45° and 90°

Size	Hours per Elbow
½ in.	.20
¾	.25
1	.30
1¼	.38
1½	.40
2	.44
2½	.59
3	.79
3½	.89
4	1.00

EMT Conduit

Size	Hours per Foot
½ in.	.02
¾	.03
1	.04
1¼	.05
1½	.06
2	.07
2½	.09
3	.11
3½	.12
4	.14

EMT Elbows — 45° and 90°

Size	Hours per Elbow
1 in.	.10
1¼	.12
1½	.15
2	.17
2½	.18
3	.20
3½	.22
4	.24

EMT Connectors — Set Screw Type

Size	Hours per Connector
½ in.	.06
¾	.06
1	.07
1¼	.08
1½	.10
2	.13
2½	.18
3	.20
3½	.22
4	.25

EMT Couplings — Set Screw Type

Size	Hours per Coupling
½ in.	.03
¾	.03
1	.04
1¼	.05
1½	.07
2	.10
2½	.15
3	.17
3½	.19
4	.21

EMT Connectors — Compression Type

Size	Hours per Connector
½ in.	.08
¾	.08
1	.09
1¼	.10
1½	.12
2	.15
2½	.23
3	.25
3½	.27
4	.30

INSTALLATION LABOR UNITS *(cont.)*

EMT Couplings — Compression Type

Size	Hours per Coupling
½ in.	.05
¾	.05
1	.06
1¼	.07
1½	.09
2	.12
2½	.20
3	.22
3½	.24
4	.27

PVC Conduit

Size	Hours per Foot
½ in.	.02
¾	.03
1	.04
1¼	.05
1½	.06
2	.07
2½	.09
3	.10
3½	.11
4	.12
5	.13
6	.14

PVC Elbows — 45° and 90°

Size	Hours per Elbow
½ in.	.10
¾	.10
1	.13
1¼	.15
1½	.18
2	.20
2½	.25
3	.30
3½	.35
4	.40
5	.50
6	.60

INSTALLATION LABOR UNITS *(cont.)*

PVC Couplings

Size	Hours per Coupling
½ in.	.05
¾	.05
1	.05
1¼	.06
1½	.06
2	.06
2½	.07
3	.07
3½	.07
4	.09
5	.10
6	.10

PVC Male and Female Adaptors

Size	Hours per Adaptor
½ in.	.10
¾	.10
1	.13
1¼	.15
1½	.16
2	.18
2½	.20
3	.25
3½	.30
4	.35
5	.40
6	.40

PVC Spacers

Size	Hours per Spacer
All sizes	.05

INSTALLATION LABOR UNITS *(cont.)*

Aluminum Rigid Conduit

Size	Hours per Foot
½ in.	.02
¾	.03
1	.04
1¼	.05
1½	.06
2	.07
2½	.08
3	.09
3½	.10
4	.13
5	.15
6	.17

ARC Elbows — 45° and 90°

Size	Hours per Elbow
½ in.	.15
¾	.20
1	.25
1¼	.30
1½	.32
2	.37
2½	.50
3	.60
3½	.70
4	.80
5	1.00
6	1.50

ARC Couplings

Size	Hours per Coupling
½ in.	.05
¾	.06
1	.07
1¼	.08
1½	.09
2	.10
2½	.11
3	.12
3½	.13
4	.14
5	.15
6	.16

INSTALLATION LABOR UNITS *(cont.)*

Nipples

Size	Hours per Nipple
½ in.	.10
¾	.10
1	.10
1¼	.15
1½	.15
2	.15
2½	.20
3	.20
3½	.20
4	.25
5	.25
6	.25

Chase Nipples

Size	Hours per Nipple
½ in.	.05
¾	.06
1	.07
1¼	.08
1½	.09
2	.10
2½	.12
3	.15
3½	.17
4	.20
5	.25
6	.30

INSTALLATION LABOR UNITS *(cont.)*

Offset Nipples

Size	Hours per Nipple
½ in.	.05
¾	.06
1	.07
1¼	.09
1½	.09
2	.10
3	.15

Pulling Elbows

Size	Hours per Elbow
½ in.	.20
¾	.24
1	.30
1¼	.40
1½	.50
2	.60

Service Head — Clamp Type

Size	Hours per Service Head
½ in.	.30
¾	.32
1	.37
1¼	.50
1½	.55
2	.63
2½	.70
3	.85
3½	1.25
4	2.10

Condulets — LB, LL, LR, C and E Types

Size	Hours per Condulet
1/2 in.	.20
3/4	.25
1	.30
1 1/4	.35
1 1/2	.37
2	.45
2 1/2	.60
3	.75
3 1/2	.90
4	1.20

Condulets — T and X Types

Size	Hours per Condulet
½ in.	.30
¾	.38
1	.45
1¼	.53
1½	.56
2	.68
2½	.90
3	1.13
3½	1.35
4	1.80

Condulet Covers

Size	Hours per Cover
½ in.	.10
¾	.10
1	.10
1¼	.12
1½	.12
2	.12
2½	.15
3	.15
3½	.15
4	.15

Seal-Off Fittings

Size	Hours per Fitting
½ in.	.50
¾	.60
1	.65
1¼	.70
1½	.75
2	.90
2½	1.40
3	1.65
4	2.00

3-Piece Couplings

Size	Hours per Coupling
½ in.	.20
¾	.22
1	.25
1¼	.30
1½	.35
2	.40
2½	.60
3	.75
3½	.90
4	1.25
5	1.65
6	2.10

INSTALLATION LABOR UNITS *(cont.)*

Bushings (Terminations)

Size	Hours per Bushing
½ in.	.10
¾	.12
1	.15
1¼	.20
1½	.22
2	.25
2½	.50
3	.65
3½	.75
4	.85
5	1.05
6	1.35

Locknuts

Size	Hours per Locknut
½ in.	.04
¾	.04
1	.05
1¼	.05
1½	.06
2	.06
2½	.07
3	.07
3½	.07
4	.08
5	.08
6	.08

Conduit Straps (including fastener)

Size	Hours per Strap
½ in.	.03
¾	.03
1	.04
1¼	.04
1½	.05
2	.06
2½	.07
3	.08
3½	.08
4	.09
5	.09
6	.10

Clamp Backs

Size	Hours per Clamp
½ in.	.05
¾	.05
1	.05
1¼	.07
1½	.07
2	.07
2½	.08
3	.08
3½	.09
4	.09

INSTALLATION LABOR UNITS *(cont.)*

Steel Boxes and Parts

Box/Parts	Hours per Item
4 in. and $4^{11}/_{16}$ in.	.15
Extension rings	.07
Blanks	.06
Plaster rings	.05
Surface covers	.07
Bar hangers	.03

Masonry Boxes

Type	Hours per Box
1 gg	.20
2 gg	.25
3 gg	.30
4 gg	.35
5 gg	.40
6 gg	.45
7 gg	.50
8 gg	.55

Cast Boxes

Size	Hours per Box
½ in.	.30
¾	.32
1	.35

Junction Boxes

Size	Hours per Box
4 x W x 4	.15
6 x W x 4	.15
8 x W x 4	.20
10 x W x 4	.25
12 x W x 4	.40
6 x W x 6	.20
8 x W x 6	.25
10 x W x 6	.40
12 x W x 6	.60
15 x W x 6	.90
18 x W x 6	1.00
24 x W x 6	1.40

Enclosures

Size/Item	Hours per Enclosure
16 x 12 x 6	.50
24 x 20 x 6	.70
30 x 24 x 6	.80
20 x 20 x 8	.80
30 x 20 x 8	.90
36 x 24 x 8	1.00
42 x 30 x 8	1.50
60 x 36 x 8	2.50
60 x 36 x 10	3.00
Plywood backing	.15

INSTALLATION LABOR UNITS *(cont.)*

Miscellaneous Items

Please note that this chart will show the time increments in various ways depending on the item. Note whether the item is by each, by the foot, by the pound, etc.

Item	Hours per Item*
Tie-wraps, small	.01
Tie-wraps, large	.02
Wire markers	.01
Wire terminals (Sta-kons)	.08
Cable supports,	
1½ – 2 in. conduit	.25
2½ – 4 in. conduit	.37
Ground clips	.02
Ground screws	.04
Grounding pigtail	.04
Duct seal, per lb	.18
Plywood backing, per 4 ft x 8 ft sheet	.75
Name plates	.10
Pitch pockets	.50
Cadwelds	.40
Fixture clips	.03

Miscellaneous Items *(cont.)*

Item	Hours per Item*
Split bolt connectors,	
No. 10 – No. 6	.15
No. 4 – 2	.30
No. 1/0 – 4/0	.50
No. 250 – 500	.80
No. 750 – 1000	1.00
Ground rods,	
8 ft	.50
10 ft	.60
Ground rod clamps,	
acorn type	.12
Grounding clamps,	
1 – 3 in. water pipe	.30
4 – 6 in. water pipe	.50
Threaded rod, per ft	.02
U channel (Unistrut),	
per ft	.03
closure strip	.01
conduit clamps,	
½ – 1 in.	.04
1¼ – 2	.06
2½ – 3½	.10
4 – 6	.14
fittings	.10

INSTALLATION LABOR UNITS *(cont.)*

Wiremold

Item	Hours per Item*
No. 200, 500, or 700	
raceway	.03
fittings	.04
No. 1500, 2000, or 2100	
raceway	.05
fittings	.11
No. 2200, 2600, 3000, or 4000	
raceway	.08
fittings	.20
No. 6000	
raceway	.11
fittings	.30
Power poles	.75
Power poles	
fittings	.15
Cable tray, 6 in.	.05
cover	.01
cross and tee fittings	1.00
other fittings	.33
Cable tray, 9 in.	.06
cover	.01
cross and tee fittings	1.10
other fittings	.37

Wiremold *(cont.)*

Item	Hours per Item*
Cable tray, 12 in.	.07
cover	.01
cross and tee fittings	1.24
other fittings	.42
Cable tray, 18 in.	.09
cover	.02
cross and tee fittings	1.40
other fittings	.48
Cable tray, 24 in.	.11
cover	.02
cross and tee fittings	1.50
other fittings	.55
Cable tray, 36 in.	.13
cover	.03
cross and tee fittings	1.75
other fittings	.60

INSTALLATION LABOR UNITS *(cont.)*

Wireway

Item	Hours per Item*
2½ x 2½ in.,	
per ft	.05
end plates and elbows	.10
4 in. x 4 in.	.06
end plates and elbows	.11
6 in. x 6 in.	.07
end plates and elbows	.12
8 in. x 8 in.	.10
end plates and elbows	.14
10 in. x 10 in.	.12
end plates and elbows	.15
12 in. x 12 in.	.15
end plates and elbows	.17
Underfloor duct, 3⅛ in.,	
per ft	.02
elbows	.25
leveling legs	.04
Underfloor duct, 7¼ in.,	
per ft	.04
elbows	.30
leveling legs	.04
Underfloor duct conduit adaptors	.15
receptacle fittings	.35
flush receptacle fittings	.50

Channeling Reinforced Concrete

Size	Hours per Foot
½ in.	.35
¾	.38
1	.40
1¼	.45
1½	.50
2	.55
2½	.65
3	.70
3½	.75
4	.90

INSTALLATION LABOR UNITS *(cont.)*

Cutting Holes in Masonry

Hole Size	Hours per Hole
$\frac{3}{4}$ in. x 8 in.	.40
$\frac{3}{4}$ in. x 12 in.	.50
$\frac{3}{4}$ in. x 16 in.	.60
$\frac{3}{4}$ in. x 20 in.	.70
$\frac{3}{4}$ in. x 24 in.	.80
$1\frac{1}{2}$ in. x 8 in.	.60
$1\frac{1}{2}$ in. x 12 in.	.70
$1\frac{1}{2}$ in. x 16 in.	.85
$1\frac{1}{2}$ in. x 20 in.	.95
$1\frac{1}{2}$ in. x 24 in.	1.10
3 in. x 8 in.	1.30
3 in. x 12 in.	1.50
3 in. x 16 in.	1.70
3 in. x 20 in.	1.90
3 in. x 24 in.	2.00
5 in. x 8 in.	1.85
5 in. x 12 in.	2.00
5 in. x 16 in.	2.20
5 in. x 20 in.	2.50
5 in. x 24 in.	2.80

INSTALLATION LABOR UNITS *(cont.)*

Sleeves

Type	Hours per Sleeve
Wall	.80
Floor	.40

Cord Strain Relief Grips

Grip Size	Hours per Grip
½ in.	.15
¾	.15
1	.15
1¼	.20
1½	.20
2	.20
2½	.35

INSTALLATION LABOR UNITS *(cont.)*

Core Drilling

Hole Size	Hours per Hole
¾ in. x 8 in.	.60
¾ in. x 12 in.	.80
¾ in. x 16 in.	1.00
¾ in. x 20 in.	1.20
¾ in. x 24 in.	1.40
1½ in. x 8 in.	.80
1½ in. x 12 in.	1.00
1½ in. x 16 in.	1.30
1½ in. x 20 in.	1.50
1½ in. x 24 in.	1.80
3 in. x 8 in.	1.70
3 in. x 12 in.	2.10
3 in. x 16 in.	2.50
3 in. x 20 in.	2.80
3 in. x 24 in.	3.30
5 in. x 8 in.	2.30
5 in. x 12 in.	2.70
5 in. x 16 in.	3.20
5 in. x 20 in.	3.80
5 in. x 24 in.	4.20

INSTALLATION LABOR UNITS *(cont.)*

Channeling Brick

Size	Hours per Foot
½ in.	.32
¾	.35
1	.38
1¼	.40
1½	.43
2	.44
2½	.50
3	.52
3½	.57
4	.60

Switchboards

Size	Hours per Switchboard
800 amp	10.00
1200 amp	16.00
1600 amp	20.00
2000 amp	26.00
3000 amp	39.00
4000 amp	52.00

INSTALLATION LABOR UNITS *(cont.)*

Distribution Panels

Size	Hours per Panel
400 amp	6.00
600 amp	8.00
800 amp	10.00
1200 amp	16.00
1600 amp	20.00
2000 amp	26.00

Motor Control Centers

Size	Hours per Center
400 amp	6.00
600 amp	8.00
800 amp	10.00
1200 amp	16.00
1600 amp	20.00
2000 amp	26.00

600-V Circuit Breakers

Size	Hours per Breaker
3-pole	
60 amp	.30
100 amp	.60
225 amp	1.20
400 amp	1.80
600 amp	3.60
800 amp	3.60
1000 amp	5.40
1200 amp	5.40

INSTALLATION LABOR UNITS *(cont.)*

Panelboards

Size/Type	Hours per Panelboard
100 amp MLO,	
12 space	3.80
20 space	5.20
100 amp MCB,	
12 space	4.50
20 space	5.70
200 amp MLO,	
24 space	7.30
40 space	8.50
200 amp MCB,	
24 space	7.80
40 space	9.00

480-V Panelboards

Size/Type	Hours per Panelboard
100 amp MLO, 18 space	6.20
MCB, 18 space	7.00
200 amp MLO, 42 space	10.40
MCB, 42 space	11.40
400 amp MLO,42 space	12.20
MCB, 42 space	13.20

Load Centers

Size/Type	Hours per Load Center
100 amp MLO, 20 space	3.80
MCB, 20 space	4.25
200 amp MLO, 40 space	5.50
MCB, 40 space	6.00

Meters

Size/Type	Hours per Meter
1 phase — 200 amp	.70
3 phase — 200 amp	.90
Per additional (meter paks)	.45
Current transformer cabinet	2.40

INSTALLATION LABOR UNITS *(cont.)*

Hubs

Size	Hours per Hub
3/4 in.	.20
1	.20
1 1/4	.25
1 1/2	.25
2	.30
2 1/2	.40
Closure	.10

Transformers — Dry Type

Size/Type	Hours per Transformer
Buck/boost	.70
15 KVA	3.20
30 KVA	6.00
45 KVA	10.00
75 KVA	12.00
112.5 KVA	14.00
150 KVA	16.00
225 KVA	20.00
300 KVA	24.00
500 KVA	28.00

INSTALLATION LABOR UNITS *(cont.)*

Transformer Vibration Isolators

Size	Hours per Isolator
15 KVA	.70
30 KVA	.70
45 KVA	.90
75 KVA	.90
112.5 KVA	1.00
150 KVA	1.00
225 KVA	1.20
300 KVA	1.50
500 KVA	1.50

INSTALLATION LABOR UNITS *(cont.)*

Motor Starters

Size	Hours per Starter
00	1.50
0	1.70
1	1.80
2	2.24
3	4.00
4	4.80

Combo Starters

Size	Hours per Starter
0	2.25
1	2.70
2	3.24
3	5.30

INSTALLATION LABOR UNITS *(cont.)*

Disconnect Switches — NEMA 1, 12 and 3R

Size/Type	Hours per Switch
2-pole	
30 amp	.60
60 amp	.75
100 amp	.90
200 amp	1.50
400 amp	3.50
600 amp	5.00
800 amp	7.50
1200 amp	12.00
3-pole	
30 amp	.70
60 amp	.90
100 amp	1.20
200 amp	1.80
400 amp	4.00
600 amp	6.00
800 amp	8.50
1200 amp	13.00
3-pole with SN	
30 amp	.80
60 amp	1.00
100 amp	1.40
200 amp	2.00
400 amp	4.50
600 amp	6.50
800 amp	9.00
1200 amp	13.50

Disconnect Switches — NEMA 4 and 5

Size/Type	Hours per Switch
2-pole	
30 amp	1.50
60 amp	2.00
100 amp	2.50
200 amp	3.00
400 amp	5.80
600 amp	11.50
3-pole	
30 amp	1.80
60 amp	2.20
100 amp	2.70
200 amp	3.50
400 amp	7.00
600 amp	13.00

Conduit Entries

Size	Hours per Entry
½ – 1 in.	.20
1¼ – 2	.40
2½ – 3½	.60
4 – 6	.80

Terminations

Wire Size	Hours per Termination
#12 – 6	.05
4 – 1	.10
1/0 – 4/0	.20
250 – 500	.30
600 – 1000	.60

INSTALLATION LABOR UNITS *(cont.)*

Lighting Contactors

Size/Type	Hours per Contactor
30 amp	
3-pole	1.20
4-pole	1.30
6-pole	1.50
60 amp	
3-pole	1.80
6-pole	2.30
100 amp	
3-pole	2.80
6-pole	3.70

Copper Feeder Bus Duct

Size	Hours per Foot
800 amp	.25
1000 amp	.30
1200 amp	.35
1350 amp	.40
1600 amp	.45
2000 amp	.50
2500 amp	.55
3000 amp	.60
4000 amp	.80
5000 amp	1.00

Copper Plug-In Bus Duct

Size	Hours per Foot
225 amp	.20
400 amp	.25
600 amp	.30
800 amp	.35
1000 amp	.40
1200 amp	.45
1350 amp	.50
1600 amp	.55
2000 amp	.60
2500 amp	.70
3000 amp	.90

Copper Bus Duct Elbows

Size	Hours per Elbow
225 – 600 amp	1.20
800 – 1600 amp	1.35
2000 – 3000 amp	1.45
4000 – 5000 amp	1.70

Copper Bus Duct End Closures

Size	Hours per End Closure
225 – 600 amp	.40
800 – 1600 amp	.50
2000 – 3000 amp	.60
4000 – 5000 amp	.80

Copper Bus Duct Tap Boxes

Size	Hours per Tap Box
225 amp	2.20
400 amp	2.80
600 amp	4.20
800 amp	4.40
1000 amp	5.70
1200 amp	5.90
1350 amp	7.20
1600 amp	7.40
2000 amp	9.00
2500 amp	12.00
3000 amp	15.00
4000 amp	19.00
5000 amp	24.00

Copper Bus Duct Circuit Breaker Adaptors

Size	Hours per Adaptor
225 – 600 amp	1.20
800 – 1600 amp	1.40
2000 amp	1.50

Aluminum Feeder Bus Duct

Size	Hours per Foot
600 – 1000 amp	.20
1200 – 1600 amp	.25
2000 amp	.30
2500 amp	.40
3000 amp	.50
4000 amp	.60

Aluminum Plug-In Bus Duct

Size	Hours per Foot
225 – 600 amp	.20
800 – 1000 amp	.25
1200 amp	.30
1350 amp	.35
1600 amp	.45
2000 amp	.50
2500 amp	.55
3000 amp	.60
4000 amp	.70

Aluminum Bus Duct Elbows

Size	Hours per Elbow
225 – 600 amp	1.00
800 – 1600 amp	1.20
2000 – 3000 amp	1.30
4000 amp	1.40

Aluminum Bus Duct End Closures

Size	Hours per End Closure
225 – 600 amp	.30
800 – 1600 amp	.40
2000 – 3000 amp	.50
4000 amp	.70

INSTALLATION LABOR UNITS *(cont.)*

Aluminum Bus Duct Tap Boxes

Size	Hours per Tap Box
225 amp	2.00
400 amp	2.60
600 amp	4.00
800 amp	4.50
1000 amp	5.50
1200 amp	6.00
1350 amp	7.00
1600 amp	7.50
2000 amp	9.00
2500 amp	11.50
3000 amp	14.50
4000 amp	19.00

Aluminum Bus Duct Circuit Breaker Adaptors

Size	Hours per Adaptor
225 – 600 amp	1.00
800 – 1600 amp	1.20
2000 amp	1.30

Miscellaneous Distribution and Control Items

Item	Hours per Item
Fuses,	
30 – 200 amp	.05
200 – 400 amp	.08
400 – 600 amp	.11
600 – 800 amp	.20
800 – 1200 amp	.30
Auxiliary contacts	.44
H-O-A switches	.50
Pilot lights	.40
Control stations, 2-button	.54
Legend plates	.06
Control relay,	
2-pole	.70
3-pole	.90
4-pole	1.10
6-pole	1.50
10-pole	2.25
Relay mounting track,	
18 in.	.20
36 in.	.30
Motor toggle switches,	
2-pole, with NEMA 1 enclosure	.65
3-pole, with NEMA 1 enclosure	.75

Miscellaneous Distribution and Control Items *(cont.)*

Item	Hours per Item
Thermal overloads	.06
Circuit breakers, 240-V,	
1-pole, 15 – 60 amp	.15
2-pole, 15 – 60 amp	.20
70 – 125 amp	.30
130 – 200 amp	.40
3-pole, 15 – 60 amp	.25
70 – 125 amp	.35
130 – 200 amp	.45
Circuit breakers, 600-V,	
2-pole, 15 – 60 amp	.30
3-pole, 15 – 60 amp	.34
70 – 100 amp	.50
100 – 225 amp	.70
250 – 400 amp	.90
450 – 800 amp	2.60
Plug-in relay with base, 8-pole	.60

INSTALLATION LABOR UNITS *(cont.)*

Time Clocks

Size/Type	Hours per Time Clock
1-pole, 24 hr.	.60
3-pole, 24 hr.	.80
3-pole, 7 day	.94
3-pole, 7 day with reserve	1.10
Astronomic dial with reserve	1.20

Photo Cells

Cell Height	Hours per Cell
Mounted at 12 ft	.45
Mounted at 20 ft	.70

INSTALLATION LABOR UNITS *(cont.)*

Copper Wire — Types THHN, THW, XHHW, etc.

Size	Hours per Foot
No. 14 AWG	.004
12	.005
10	.006
8	.007
6	.008
4	.009
3	.011
2	.012
1	.014
1/0	.018
2/0	.020
3/0	.025
4/0	.028
250 KCMIL	.029
300	.032
350	.039
400	.046
500	.053
600	.060
750	.067
1000	.077
Pull wire	.004
No. 18 AWG	.003
No. 16	.003

Aluminum Wire — Types THHN, THW, XHHW, etc.

Size	Hours per Foot
No. 6 AWG	.007
4	.008
3	.009
2	.011
1	.013
1/0	.015
2/0	.018
3/0	.021
4/0	.023
250 KCMIL	.025
300	.027
350	.031
400	.037
500	.043
600	.050
750	.056
1000	.065

Low Voltage Cables

Size/Type	Hours per Foot
Coaxial cable	.005
Thermostat cable,	
4-conductor	.005
8-conductor	.010
Control cables,	
4-conductor	.005
8-conductor	.009
12-conductor	.013
20-conductor	.023
30-conductor	.025
50-conductor	.031
Telephone cables,	
2-pair	.005
6-pair	.010
10-pair	.018
25-pair	.032

INSTALLATION LABOR UNITS *(cont.)*

5000-V Copper Wire

Size	Hours per Foot
No. 8 AWG	.009
6	.010
4	.013
2	.015
1	.018
1/0	.023
2/0	.028
3/0	.030
4/0	.033
250 KCMIL	.038
350	.048
500	.065

15,000-V Copper Wire

Size	Hours per Foot
No. 4 AWG	.018
2	.023
1	.028
1/0	.030
2/0	.033
3/0	.038
4/0	.043
250 KCMIL	.050
350	.061
500	.078

INSTALLATION LABOR UNITS *(cont.)*

Building Wire Cables

Size/Type	Hours per Foot
Type NM No. 14 or No. 12,	
2-conductor with ground	.006
3-conductor with ground	.008
Type NM No. 10,	
2-conductor with ground	.008
Type NM No. 6,	
2-conductor with ground	.011
Type SEU No. 4, 3-conductor	.016
3	.018
2	.020
1	.024
1/0	.027
2/0	.033
3/0	.040
4/0	.047
Type AC (BX) cable,	
No. 14, 2-conductor	.012
3-conductor	.014
4-conductor	.016
Type AC (BX) cable,	
No. 12, 2-conductor	.016
3-conductor	.018
4-conductor	.021
No. 10, 2-conductor	.018
3-conductor	.021

INSTALLATION LABOR UNITS *(cont.)*

Light Fixtures (lamp included)

Size/Type	Hours per Fixture
2 ft x 4 ft recessed fluorescent	.70
2 ft x 2 ft	.65
1 ft x 4 ft	.60
8 ft, 2-lamp strip	.70
8 ft, 1-lamp strip	.65
4 ft, 2-lamp strip	.60
4 ft, 1-lamp strip	.57
8 ft, 2-lamp wraparound	.85
4 ft, 2-lamp wraparound	.75
175-W recessed HID	1.00
250-W	1.20
400-W	1.50
175-W open HID	1.40
250-W	1.80
400-W	2.20
175-W wall pack	1.00
150-W recessed incandescent	.75
300-W	.90

INSTALLATION LABOR UNITS *(cont.)*

Light Fixtures (lamp included) *(cont.)*

Size/Type	Hours per Fixture
100-W surface incandescent	.50
Over-mirror light	.55
Exit sign	.50
Recessed exit sign	.80
Emergency ballast	.40
2-head emergency light	.80
Bollard	1.40
25 ft pole with 2 heads	4.00
Bath exhaust fan	.54
4 ft track	.90
8 ft track	1.00
12 ft track	1.20
Track lighting head	.21
L or T	.20
Porcelain fixture	.24
Explosionproof incandescent	1.30
Explosionproof fluorescent, 4 ft, 2-lamp	2.00

INSTALLATION LABOR UNITS *(cont.)*

Wiring Devices

Size/Type	Hours per Device
Duplex receptacle	.17
240-V receptacle,	
20 amp	.20
30 amp	.30
50 amp	.42
GFI receptacle	.24
1-pole switch	.17
3-way switch	.21
4-way switch	.24
1-pole switch with pilot light	.20

Trim Plates

Type	Hours per Plate
1 gg	.05
2 gg	.07
3 gg	.09
4 gg	.11
Weatherproof	.08

INSTALLATION LABOR UNITS *(cont.)*

Greenfield and Sealtite

Size	Hours per Foot
½ in.	.02
¾	.02
1	.03
1¼	.04
1½	.05
2	.06
2½	.08
3	.10
3½	.11
4	.12

Greenfield Connectors — Straight

Size	Hours per Connector
½ in.	.07
¾	.07
1	.09
1¼	.10
1½	.11
2	.12
2½	.15
3	.17
3½	.20
4	.25

INSTALLATION LABOR UNITS *(cont.)*

Greenfield Connectors — 90°

Size	Hours per Connector
½ in.	.10
¾	.10
1	.12
1¼	.13
1½	.14
2	.15
2½	.18
3	.20
3½	.25
4	.30

Sealtite Connectors — Straight

Size	Hours per Connector
½ in.	.10
¾	.10
1	.12
1¼	.15
1½	.16
2	.17
2½	.20
3	.25
4	.30

Sealtite Connectors — 90°

Size	Hours per Connector
½ in.	.12
¾	.12
1	.14
1¼	.17
1½	.18
2	.19
2½	.22
3	.27
4	.32

Signaling Systems

Item	Hours per Item
Bell	.40
Buzzer	.30
Horn	.40
Chime	.40
Push button	.30
Door opener	.25
Transformer	.30

INSTALLATION LABOR UNITS *(cont.)*

Emergency Generators

Size	Hours per Generator
10.0 kw	10.00
15.0	12.00
30.0	18.00
50.0	22.00
75.0	24.00
100.0	30.00
125.0	36.00
150.0	38.00
175.0	40.00
200.0	50.00
250.0	54.00
300.0	62.00
350.0	74.00
400.0	84.00
500.0	98.00

Automatic Transfer Switches — 3-Pole

Size	Hours per Switch
70 amp	2.90
100 amp	3.80
150 amp	4.70
260 amp	7.00
400 amp	7.00
600 amp	11.50
800 amp	11.50
1000 amp	16.00
1200 amp	16.00
1600 amp	20.50
2000 amp	25.00

INSTALLATION LABOR UNITS *(cont.)*

Flat Conductor Cable and Parts

Cable Size/Item	Hours per Foot/Item
3-conductor cable	.04
4-conductor cable	.05
5-conductor cable	.06
10-conductor communications cable	.04
16-conductor communications cable	.04
24-conductor communications cable	.05
Duplex receptacle service heads	.60 each
Double duplex receptacle service heads	.70
Telephone service heads	.60
Receptacle and telephone service heads	.70
Blank cover service heads	.05
Surface transition box	.40
Recessed transition box	.60
End caps	.05
Insulators	.15
Splice connectors	.30
Tap connectors	.30
Cable connectors	.30
Terminal blocks	.40

Heaters and Heating Devices

Item/Size/Type	Hours per Device
Baseboard heaters	
500 – 750 W	.50
800 – 1250	.65
1300 – 2000	.75
2100 – 2500	1.00
Thermostats	.40
Corner and blank sections	.20
Receptacle sections	.40
Wall heaters,	
750 – 2000 W	1.00
2500 – 4000	1.40
Infrared heaters,	
600 – 2000 W	.70
2500 – 4000	1.00
Radiant ceiling panels	.50
Radiant ceiling panels frame	.20

Residential (Romex) Cable

Size/Type	Hours per Foot
No. 14/2 cable	.005
14/3	.006
14/4	.008
12/2	.006
12/3	.007
12/4	.009
10/2	.007
10/3	.008
8/2	.009
6/2	.01
No. 6/2 aluminum	.008

INSTALLATION LABOR UNITS *(cont.)*

Residential Items and Devices

Item/Device	Hours per Item/Device
Nail-on box	.05
Ceiling box	.07
Plastic finish plate	.04
Weatherproof plate	.06
Receptacle	.06
GFI receptacle	.12
Flush dryer outlet	.12
1-pole switch	.05
3-way switch	.06
Residential light fixture (simple)	.25
Recessed fixture	.40
Exhaust fan	.30
Smoke detector, 120 V	.20
Surface range outlet	.20
Door chime	.20
Push button	.07
Coaxial cable	.004
Telephone cable	.004

CHAPTER 2
Unit Pricing

CONCEPTS OF UNIT PRICING

The concept of unit pricing is very simple. Unit pricing establishes a set dollar amount to charge for each particular item or group of items identified by one symbol or name. For example, a customer will ask for the price of each receptacle. However, he doesn't want the price for the duplex receptacle alone, but the price for the receptacle, trim plate, box, ring, conduit or cable, and labor. He wants a total price for each particular item that he can see and use, and he identifies it by one name — a receptacle.

Because of the convenience of unit pricing, it has gained a large degree of acceptance. Using a good set of unit prices, an estimator can price a complete job by just counting the symbols on the plans and multiplying them by their unit prices to determine the selling price. This eliminates the long and difficult task of making an itemized, detailed estimate for each job.

The most important concept to understand about unit pricing is that a unit price is your price for *everything* required, whether specifically expressed or not.

For example, we understand that the duplex receptacle unit contains items such as the box, trim plate, fasteners, and raceway or cable, but what about the circuit breaker that must feed power to the receptacle? This must also be included in the unit if the price is to be accurate.

CONCEPTS OF UNIT PRICING *(cont.)*

The units shown in the following figures were specifically developed for commercial work (**Figures 2.1 to 2.10**) and residential work (**Figures 2.11 to 2.26**) and include everything that pertains to them. If we can remember to include *every* item that goes into making up the unit, we then have a *complete* unit that we can price and use with full confidence.

The 30 percent O & P for commercial work shown will yield 23.1 percent of your total sales toward overhead and profit, an average rate.

The 50 percent O & P shown for residential work will yield 33.3 percent of your total sales toward overhead and profit, and this percentage has proven workable for residential work.

To construct many of these units, assemblies are used. This is to make the task easier and faster. By taking a similar unit and then modifying it, we can save the time and effort of listing, pricing, and extending each individual part.

There are certain parts of these estimates that you should be aware of:

1. Always included is the cost of material and labor for an appropriate fastener with each strap. This ensures a sufficient number of fasteners of the correct type and does so without using a separate process.

2. Also included is some miscellaneous material and labor in each unit for such things as tape, wire lube, wire markers and testing.

3. The material cost of light fixtures or fans in the residential units is omitted, as most home builders like to furnish these items.

4. The price and labor for No. 14 Romex (Type NM Cable) is a 4 to 1 average ratio of No. 14/2 to No. 14/3.

5. No double switches are shown for residential units because there is no significant difference between two switches and a double switch. A two-gang plastic box costs just about twice as much as a one-gang box, as does the trim plate, and there is no difference in the amount of cable. For the commercial units, double switches are included because there is a substantial savings in raceway.

Be sure to include the cost of the permit or any other unusual expenses when you total your estimate, and watch for any exceptionally long runs or especially heavy equipment, which must be compensated for.

Once you understand the concepts and use of unit pricing, you should be able to develop unit prices for any needs you have with confidence and assurance. Keep in mind that you will have to price according to your local conditions.

FIGURE 2.1 — DUPLEX RECEPTACLE

COMPLETED ESTIMATE SHEET

Job ______________________ Page ______

______________________ of ______ Pages

Estimated by ______________ Checked by ______________ Date ______________

Description	Material				Labor		
	Quantity	Unit Price	Per	Amount	Unit	Per	Amount
Duplex Receptacle							
Receptacle	1	50	E	50	.17	E	17
Plate	1	19	E	19	.05	E	05
1-gg, 1/2" Raise Ring	1	40	E	40	.05	E	05
4" Square Box	1	54	E	54	.17	E	17
1/2" EMT	20'	13 00	C	2 60	.02	E	40
Connectors	2	09	E	18	.06	E	12
Cplg	1 1/2	09	E	14	.03	E	05
Strap with Fastener	2 1/2	09	E	23	.03	E	08
Fasteners	3	04	E	12	.01	E	03

Item	Quantity	Material Price	Per	Material Extension	Labor	Per	Labor Extension
#12 THHN Wire	46'	30 00	M	1 38	.006	E	12
Grounding Pigtail	1	22	E	22	.07	E	07
Wire Nuts	2	05	E	10	—	—	—
Circuit Breaker	1/6	4 00	E	67	.12	E	02
Misc., Tape, Wire Lube, etc.	—	—	—	11	—	—	01
				7 38			1 34

Material	7.38
Tax	.37
	7.75
1.34 hr @ 11.81	15.83
	23.58
30% O & P	7.07
	$30.65

Example of a completed unit pricing sheet for a standard, duplex receptacle.

FIGURE 2.2 — DUPLEX RECEPTACLE

ESTIMATE SHEET

Job ____________________ **Page** ________

____________________ **of** ________ **Pages**

Estimated by ____________ **Checked by** ____________ **Date** ________

Description	Material				Labor		
	Quantity	Unit Price	Per	Amount	Unit	Per	Amount
Duplex Receptacle							
Receptacle	1						
Plate	1						
1-gg, 1/2" Raise Ring	1						
4" Square Box	1						
1/2" EMT	20'						
Connectors	2						
Cplg	1 1/2						
Strap with Fastener	2 1/2						
Fasteners	3						

Item	Quantity
#12 THHN Wire	46'
Grounding Pigtail	1
Wire Nuts	2
Circuit Breaker	1/6
Tape, etc.	—
Material	
Tax	
hr @	
30% O & P	
$	

The duplex receptacle consists of the device, trim plate, 1-gang (gg) plaster ring, box, grounding pigtail, conduit and wire, fasteners, a portion of a circuit breaker, wire nuts, tape, and so on.

FIGURE 2.3 — SINGLE-POLE SWITCH

ESTIMATE SHEET

Job __________ Page ______

of ______ Pages

Estimated by ______ Checked by ______ Date ______

Description	Material				Labor		
	Quantity	Unit Price	Per	Amount	Unit	Per	Amount
Single-Pole Switch							
1-Pole Switch	1						
Plate	1						
1-gg, 1/2" Raise Ring	1						
4" Square Box	1						
1/2" EMT	16'						
Connectors	2						
Cplg	1 1/2						
Strap	2						
Fasteners	3						
#12 THHN Wire	38'						

Wire Nuts	2
Misc., Tape, Wire Lube, etc.	—
Material	
Tax	
hr @	
30% O & P	
$	

The single-pole switch unit consists of a 16-foot run to the switch and its associated items. Note that this unit does not include a portion of a circuit breaker, as it is not a current consuming device.

FIGURE 2.4 — THREE-WAY SWITCH

ESTIMATE SHEET

Job ______________________ **Page** ______

______________________ **of** ______ **Pages**

____

____

Estimated by ______________ **Checked by** ______________ **Date** ______

Description	Material				Labor		
	Quantity	Unit Price	Per	Amount	Unit	Per	Amount
Three-Way Switch							
3-Way Switch	1						
Plate	1						
1-gg, 1/2" Raise Ring	1						
4" Square Box	1						
1/2" EMT	20'						
Connectors	2						
Cplg	1 1/2						
Strap	2 1/2						
Fasteners	3						
#12 THHN Wire	100'						

Item	Qty
Wire Nuts	4
Tape, etc.	—
Material	
Tax	
hr @	
30% O & P	
$	

The three-way switch unit is similar to the single-pole switch unit, but has a slightly longer run. Note also that there is an increased amount of wire in this unit. This is because three-way switches are usually located a good distance apart on commercial jobs, and the raceway for the light fixtures is used to run a portion of the wiring for the switches.

FIGURE 2.5 — TWO-BY-FOUR LAY-IN FIXTURE

ESTIMATE SHEET

Job ______________________________ Page ________

______________________________ of ________ Pages

Estimated by ____________ Checked by ____________ Date ________

Description	Material				Labor		
	Quantity	Unit Price	Per	Amount	Unit	Per	Amount
Two-by-Four Lay-In Fixture (Four Lamps)							
2 x 4 Fixture, Prelamped,							
Prewhipped	1						
Fixture Clips	2						
4" Square Box	1						
Blank Cover	1						
Fasteners	2						
1/2" EMT	12'						
Connectors	2						
Cplg	1						

Item	Quantity
Strap	1 1/2
#12 THHN Wire	40'
Wire Nuts	3
Circuit Breaker	1/6
Tape, etc.	—
Material	
Tax	
hr @	
30% O & P	
$	

The 2 x 4 fixture unit consists of a standard 2 ft. by 4 ft., four-lamp, recessed fluorescent light fixture with branch circuitry and associated items. The estimate shows a prelamped, prewhipped fixture.

FIGURE 2.6 — BATHROOM LIGHT

ESTIMATE SHEET

Job ______________________________ Page ________

______________________________ of ________ Pages

Estimated by ____________ Checked by ____________ Date ________

Description	Material				Labor		
	Quantity	Unit Price	Per	Amount	Unit	Per	Amount
Bathroom Light							
Wall-Mounted Incandescent							
Fixtures with Lamp	1						
4" Square Box	1						
Round Ring	1						
1/2" EMT	15'						
Connectors	2						
Cplg	1 1/2						
Strap	2						
Fasteners	3						

Item	Quantity
#12 THHN Wire	36'
Wire Nuts	3
Circuit Breaker	1/6
Tape, etc.	—
Material	
Tax	
hr @	
30% O & P	
$	

The wall-mounted fixture unit consists of a simple incandescent fixture with lamp mounted on a wall outlet. Included is the branch circuitry and associated items.

FIGURE 2.7 — EMERGENCY LIGHT

ESTIMATE SHEET

Job ______________________ Page ______

______________________ of ______ Pages

Estimated by ______________ Checked by ______________ Date ______

Description	Material				Labor		
	Quantity	Unit Price	Per	Amount	Unit	Per	Amount
Emergency Light							
2-Head Emergency Light	1						
4" Square Box	1						
1-gg, 1/2" Raise Ring	1						
Fasteners	3						
1/2" EMT	16'						
Connectors	2						
Cplg	1 1/2						
Strap	2						
#12 THHN Wire	39'						

Wire Nuts	2
Circuit Breaker	1/10
Tape, etc.	—
Material	
Tax	
hr @	
30% O & P	
$	

The emergency light unit consists of a two-head emergency light mounted over a wall outlet. Branch circuitry and associated items are included.

FIGURE 2.8 — PHONE STUB

ESTIMATE SHEET

Job ______________________________ Page ________

______________________________ of ________ Pages

Estimated by ____________ Checked by ____________ Date ________

Description	Material				Labor		
	Quantity	Unit Price	Per	Amount	Unit	Per	Amount
Phone Stub							
$1/2$" EMT	10'						
Connectors	1						
Strap	1						
4" Square Box	1						
1-gg, $1/2$" Raise Ring	1						
Fasteners	3						
Tape, etc.	—						
Material							

Tax

hr @

30% O & P

$

The phone stub unit is a stub-out from a box in the wall to above the suspended ceiling. The telephone contractor will complete the installation.

FIGURE 2.9 — THREE-PHASE ROOFTOP UNIT

ESTIMATE SHEET

Job ______________________ **Page** ______

______________________ **of** ______ **Pages**

Estimated by ______________ **Checked by** ______________ **Date** ______

Description	Material				Labor		
	Quantity	Unit Price	Per	Amount	Unit	Per	Amount
Three-Phase Rooftop Unit							
60 Amp, 3-Pole,							
Nonfused Disconnect	1						
3/4" Chase Nipple	1						
Locknut	1						
3/4" Sealtite	4'						
Conn-Straight	2						
Pipe Coupling	1						
Fasteners	3						
Terminations	4						

Wire Nuts	4
$3/4$" EMT	50'
Connectors	1
Conn-Compression	1
Cplg	5
Strap	6
#6 THHN Wire	186'
#10 THHN Wire	62'
3-Pole Circuit Breaker	1
Tape, etc.	—
Material	
Tax	
hr @	
30% O & P	
$	

The three-phase rooftop unit consists of a 60 amp, three-phase circuit from the panel to a rooftop heating, ventilating and air-conditioning unit. A disconnect is mounted on the side of the unit, and the last four feet of raceway to the disconnect is Sealtite. Also included are terminations, wire nuts, fasteners, fittings, a 60 amp three-pole circuit breaker, etc. Note that any required pitch pan is to be installed by another contractor.

FIGURE 2.10 — 200 AMP, THREE-PHASE OVERHEAD SERVICE

ESTIMATE SHEET

Job ______________________________ Page ________

______________________________ of ________ Pages

____

____

Estimated by ____________ Checked by ____________ Date ________

Description	Material				Labor		
	Quantity	Unit Price	Per	Amount	Unit	Per	Amount
200 Amp, Three-Phase Overhead Service							
$2^1/_2$" GRC	10'						
Head	1						
Back Plates	2						
Straps	2						
Hub	1						
Nipple	1						
Locknuts	4						
Bushings	2						
Meter-Furnished by Utility	1						
MCB Panel—42 Space	1						
#250 AL THW Wire	60'						

Item	Quantity
#2 AL THW Wire	20'
$1/2$" EMT	30'
Connectors	3
Cplg	3
Strap	4
Ground Rod	1
Clamp	1
Water Pipe Clamp	1
#4 Bare Copper Wire	45'
Fasteners	18
Cut and Patch Wall	1
Tape, etc.	—
Material	
Tax	
hr @	
30% O & P	
$	

The 200 amp, three-phase overhead service unit itemizes a basic service. The panel is a load center; the grounding electrode conductor is taken to both the ground rod and to a water pipe in $1/2$ inch EMT. The service conductors are aluminum, sized as per the NEC.

FIGURE 2.11 — DUPLEX RECEPTACLE

ESTIMATE SHEET

Job ____________________ Page ______

____________________ of ______ Pages

Estimated by __________ Checked by __________ Date ______

Description	Material				Labor		
	Quantity	Unit Price	Per	Amount	Unit	Per	Amount
Duplex Receptacle							
Plastic Nail-On Box	1						
Duplex Receptacle	1						
Plastic Plate	1						
#14 Cable	20'						
Circuit Breaker	$^1/_6$						
Staples, Wire Nuts, etc.	—						
Material							
Tax							

hr @	
30% O & P	
$	

The duplex receptacle unit consists of the device, trim plate, box, 20 ft. of #14 cable, wire nuts, staples, and a portion of a circuit breaker. We add the associated labor, overhead, and profit.

FIGURE 2.12 — GROUND-FAULT INTERRUPTER RECEPTACLE

ESTIMATE SHEET

Job ______________________________ Page ________

______________________________ of ________ Pages

____

____

Estimated by ______________ Checked by ______________ Date ________

Description	Material				Labor		
	Quantity	Unit Price	Per	Amount	Unit	Per	Amount
Ground-Fault Interrupter Receptacle							
Duplex Receptacle Assembly	1						
Duplex — Deduct	1						
Plate — Deduct Material Only	1						
GFI	1						
Misc., Wire Nuts, etc.	—						
Material							
Tax							

hr @

50% O & P

$

For the GFI unit, we begin with the duplex receptacle unit, then deduct the plate and device, and add the GFI receptacle, which comes with its own finish plate. We add a little extra for the extra wire nuts, which are required.

FIGURE 2.13 — 20 AMP OUTLET

ESTIMATE SHEET

Job ______________________________ **Page** ________

______________________________ **of** ________ **Pages**

____

____

Estimated by ____________ **Checked by** ____________ **Date** ________

Description	Material				Labor		
	Quantity	Unit Price	Per	Amount	Unit	Per	Amount
20 Amp Outlet (Washer, Freezer, etc.)							
Plastic Box	1						
Receptacle	1						
Plate	1						
#12/2 Cable	24'						
Circuit Breaker	1						
Staples, etc.	—						
Material							
Tax							

hr @	
50% O & P	
$	

The 20 amp circuit serves one receptacle, which is typical for a washer, microwave or a freezer. It is similar to the duplex receptacle, but uses #12/2 cable, and has its own circuit breaker.

FIGURE 2.14 — DRYER OUTLET

ESTIMATE SHEET

Job ______________________________ **Page** __________

______________________________ **of** __________ **Pages**

____

____

Estimated by __________ **Checked by** __________ **Date** __________

Description	Material				Labor		
	Quantity	Unit Price	Per	Amount	Unit	Per	Amount
Dryer Outlet							
Plastic Box	1						
Dryer Receptacle	1						
Plate	1						
#10/2 Cable	24'						
2-Pole Circuit Breaker	1						
Staples, etc.	—						
Material							
Tax							

hr @

50% O & P

$

The dryer outlet unit consists of a 24 ft. run of #10/2 cable from the panel to a flush mounted dryer receptacle. It includes the device, plate, box, circuit breaker and staples.

FIGURE 2.15 — WATER HEATER

ESTIMATE SHEET

Job ____________________ **Page** ______

____________________ **of** ______ **Pages**

Estimated by ______ **Checked by** ______ **Date** ______

Description	Material				Labor		
	Quantity	Unit Price	Per	Amount	Unit	Per	Amount
Water Heater							
#10/2 Cable	24'						
2-Pole Circuit Breaker	1						
Cable Connector	1						
Terminations	3						
Staples, etc.	—						
Material							
Tax							

hr @

50% O & P

$

The water heater unit consists of a 24 ft. run of #10/2 cable from the panel to an electric water heater. It includes the required circuit breaker, a cable connector, labor for terminating the wires and miscellaneous items.

FIGURE 2.16 — AIR CONDITIONING UNIT

ESTIMATE SHEET

Job ______________________________ Page ________

of ________ Pages

Estimated by ______________ Checked by ______________ Date ________

Description	Material				Labor		
	Quantity	Unit Price	Per	Amount	Unit	Per	Amount
Air Conditioning Unit							
#10/2 Cable	30'						
Disconnect Switch — WP	1						
1/2" Sealtite	4'						
Connectors-Straight	2						
#10 THHN Wire	18'						
2-Pole Circuit Breaker	1						
Terminations	9						
Staples, Wire Nuts, etc.	—						

Material	
Tax	
hr @	
50% O & P	
$	

The air conditioning unit consists of a 30 ft. run of #10/2 cable from the panel to an outdoor disconnect switch, then to an air conditioning unit in a 4 ft. run of $^1/_2$ inch Sealtite. It includes a two-pole circuit breaker and labor for terminating the wires in the unit, screws and staples.

FIGURE 2.17 — AIR HANDLING UNIT/HEATER

ESTIMATE SHEET

Job ______________________________ Page ________

______________________________ of ________ Pages

____

____

Estimated by ____________ Checked by ____________ Date ________

Description	Material				Labor		
	Quantity	Unit Price	Per	Amount	Unit	Per	Amount
Air Handling Unit/Heater							
#6/2 Cable	22'						
Disconnect Switch	1						
Cable Connectors	3						
2-Pole Circuit Breaker	1						
Terminations	9						
Staples, Screws, etc.	—						
Material							
Tax							

hr @

50% O & P

$

The air handling unit/heater unit is a 20 ft. run of #6/2 cable from the panel to a disconnect switch mounted next to the unit, then a short run of cable into the unit. It includes the required circuit breaker and labor for terminating the wires, cable connectors, etc.

FIGURE 2.18 — RANGE OUTLET

ESTIMATE SHEET

Job ______________________ Page ______

______________________ of ______ Pages

____

____

Estimated by ______________ Checked by ______________ Date ______

Description	Material				Labor		
	Quantity	Unit Price	Per	Amount	Unit	Per	Amount
Range Outlet							
Surface Receptacle	1						
#6/2 Cable	24'						
2-Pole Circuit Breaker	1						
Staples, Screws, etc.	—						
Material							
Tax							
hr @							

50% O & P

$

The range outlet unit consists of a 24 ft. run of #6/2 cable from the panel to a surface mounted range receptacle and associated items.

FIGURE 2.19 — SINGLE-POLE SWITCH

ESTIMATE SHEET

Job ______________________________ Page ________

______________________________ of ________ Pages

____

____

Estimated by ______________ Checked by ______________ Date ________

Description	Material				Labor		
	Quantity	Unit Price	Per	Amount	Unit	Per	Amount
Single-Pole Switch							
Plastic Box	1						
1-Pole Switch	1						
Plate	1						
#14 Romex	15'						
Staples, etc.	—						
Material							
Tax							

hr @

50% O & P

$

The single-pole switch unit includes the device, box, trim plate, a 15 ft. average run of #14 cable, staples, and associated labor, overhead and profit.

FIGURE 2.20 — LIGHT FIXTURE OUTLET

ESTIMATE SHEET

Job ______________________________ **Page** ________

______________________________ **of** ________ **Pages**

____

____

Estimated by ____________ **Checked by** ____________ **Date** ____________

Description	Material				Labor		
	Quantity	Unit Price	Per	Amount	Unit	Per	Amount
Light Fixture Outlet							
#14 Romex	15'						
Round Plastic Box	1						
Circuit Breaker	$^1/_6$						
Fixture and Lamp	1						
Staples, Wire Nuts, etc.	—						
Material							
Tax							

hr @																		
50% O & P																		
$																		

The light fixture unit consists of a 15 ft. run of #14 cable, a round plastic box, a portion of a circuit breaker (as it is a current consuming device), labor for installing the simple fixture, and so on.

FIGURE 2.21 — BATH EXHAUST FAN

ESTIMATE SHEET

Job ______________________ Page ______

______________________ of ______ Pages

____

____

Estimated by ______________ Checked by ______________ Date ______

Description	Material				Labor		
	Quantity	Unit Price	Per	Amount	Unit	Per	Amount
Bath Exhaust Fan							
Exhaust Fan	1						
#14 Romex	16'						
Terminations	3						
Circuit Breaker	$^{1}/_{10}$						
Cable Connector	1						
Staples, etc.	—						
Material							
Tax							

hr @																
50% O & P																
$																

The bath exhaust fan outlet includes a 16 ft. run of cable from the switch to the exhaust fan, labor for installing the fan and terminating the wires; also the associated items and markup.

FIGURE 2.22 — DOUBLE FLOODLIGHT

ESTIMATE SHEET

Job ______________________ Page ______

______________________ of ______ Pages

Estimated by ______ Checked by ______ Date ______

Description	Material				Labor		
	Quantity	Unit Price	Per	Amount	Unit	Per	Amount
Double Floodlight							
#14 Romex	30'						
Box	1						
Cluster Base	1						
Lamp Holder	2						
PAR Lamp	2						
Circuit Breaker	1/4						
Staples, Wire Nuts, etc.	—						
Material							

Tax

hr @

50% O & P

$

The double floodlight unit consists of a 30 ft. run to a flush box in a soffit or high on the exterior wall, onto which is mounted a cluster base, two lamp-holders, and two PAR lamps. Also included is a portion of the circuit breaker, etc.

FIGURE 2.23 — WELL PUMP

ESTIMATE SHEET

Job ____________________ Page ______

____________________ of ______ Pages

____

____

Estimated by ______ Checked by ______ Date ______

Description	Material				Labor		
	Quantity	Unit Price	Per	Amount	Unit	Per	Amount
Well Pump							
#14 Romex	20'						
#14 UF Cable	50'						
Trenching	42'						
WP Box	1						
2-Pole Switch	1						
WP Switch Cover	1						
$1/2$" PVC Conduit	5'						
90° Ell	1						
TA	1						

Item	Qty
1/2" Sealtite	5'
Connector-Straight	1
Strap	1
Terminations	3
2-Pole Circuit Breaker	1
Wire Nuts, etc.	—
Material	
Tax	
hr @	
50% O & P	
$	

The well pump unit includes a run of #14 cable from the panel to a weatherproof 2-pole switch surface mounted on the exterior wall, where it is protected by 1/2 inch PVC conduit as it goes below grade, then runs 50 ft. to the pump, where it is protected in 1/2 inch Sealtite up to its termination in the pump control.

FIGURE 2.24 — TELEPHONE OUTLET

ESTIMATE SHEET

Job ______________________ **Page** ______

______________________ **of** ______ **Pages**

____

____

Estimated by ______________ **Checked by** ______________ **Date** ______

Description	Material				Labor		
	Quantity	Unit Price	Per	Amount	Unit	Per	Amount
Telephone Outlet							
Plastic Box	1						
Plate	1						
Phone Cable	32'						
Staples, etc.	—						
Material							
Tax							
hr @							

50% O & P
$

The telephone outlet unit includes a 32 ft. average run of telephone cable, a plastic box, trim plate, staples, and so on.

FIGURE 2.25 — DOOR CHIME AND PUSH BUTTON

ESTIMATE SHEET

Job ____________________ Page ______

____________________ of ______ Pages

Estimated by ____________ Checked by ____________ Date ______

Description	Material				Labor		
	Quantity	Unit Price	Per	Amount	Unit	Per	Amount
Door Chime and Push Button							
Chime and Transformer	1						
Push Button	1						
Phone Cable	25'						
2-Gang Box	1						
#14 Cable	15'						
Circuit Breaker	$^{1}/_{10}$						
Staples, Wire Nuts, etc.	—						
Material							

Tax

hr @

50% O & P

$

The door chime unit consists of a typical door chime mounted over a two-gang plastic box, with a 15 ft. run of cable to the nearest useable outlet for power and a 25 ft. run of phone wire to an exterior push button. Also included is the transformer, staples, wire nuts, and so on.

FIGURE 2.26 — 150 AMP OVERHEAD SERVICE

ESTIMATE SHEET

Job ____________________ Page ______

____________________ of ______ Pages

Estimated by ______ Checked by ______ Date ______

Description	Material				Labor		
	Quantity	Unit Price	Per	Amount	Unit	Per	Amount
150 amp Overhead Service							
$1\frac{1}{2}$" conduit riser	10'						
$1\frac{1}{2}$" service head	1						
$1\frac{1}{2}$" connector or termination	1						
$1\frac{1}{2}$" meter hub	1						
$1\frac{1}{2}$" straps	2						
#2/0 Al. THW wire	30'						
#1 Al. THW wire	15'						
Meter (furnished by utility)	1						
#6 bare copper wire	30'						

Item	Qty
Pipe Clamp	1
Ground Rod	1
Ground Rod Clamp	1
1/2" EMT	10'
1/2" EMT connector	1
1/2" EMT strap	1
Main Circuit Breaker	
Panel & Cover	1
#2/0 Al. SEU Cable	28'
Large Cable Connectors	2
Tape, etc.	—
Material	
Tax	
hr @	
50% O & P	
$	

The 150 amp overhead service unit brings power into the building through the riser (type of conduit determined by local code), through the meter, and then to the panel via service cable. Grounding materials are also included.

UNIT PRICING NOTES

CHAPTER 3
Technology Installations

LABOR UNITS

As we go through this section, we will be explaining the basic techniques of high tech estimating. First, however, we should detail the preliminary steps. Many of these are common with conventional estimating, but they bear repeating here.

The first step in estimating is to ascertain the overall requirements of the job being estimated. You must get a clear picture in your mind of how this job will flow; and more importantly, where will the money come from, and when. In addition, you must understand the scope of the work you are quoting a price for and exactly what will be required of you. These are primary concerns and the first considerations in any good estimate.

In order to verify that all of these factors are considered, many estimators use checklists that they review for every project. You should develop your own for such uses.

The following list contains the essential elements of job site labor. Any good cost estimate must cover these operations. The conventional method of estimating does this conveniently by assigning a single labor unit to each item.

1. Reading the plans.

2. Ordering materials.

3. Receiving and storing materials.

4. **Moving the materials to where they are needed.**
5. **Getting the proper tools.**
6. **Making measurements and laying out the work.**
7. **Installing the materials.**
8. **Testing.**
9. **Cleaning up.**
10. **Lost time and breaks.**

The conventional method of estimating is excellent where circumstances are consistent and predictable. With high tech estimating, there are other factors that come into play. For instance, we can specify a certain labor rate for programmable controller, but there may be other functions required along with the controller. Note the word *may*. Functions such as programming, software installation, testing, modifications to programs, etc. These functions may or may not be necessary.

Because of this situation (which is typical to most types of high tech equipment), a single labor unit is not sufficient to assign labor for the item. What might be applicable for one installation may be double what is appropriate in the next installation — even though it is exactly the same type of equipment. The challenge of high tech estimating is finding a method of assigning labor in either situation.

In order to solve the problems of charging high tech labor, we have two possible solutions. The first is that we establish a single variable labor unit for each item of material. For example, the labor unit for the programmable controller we mentioned would be 2.0 – 20.0. If we use this method, we can continue our use of the conventional estimating method, although the use of such a broad labor range seems certain to cause problems. Secondly, a single broadly variable labor unit gives the estimator an easy way to avoid thinking about the requirements of the installation. Rather than reviewing the programming requirements for the controller, it is far easier to say, "Well, it's a tough one, call it 16 hours." The estimator may be skilled, but if he or she does not intelligently calculate all of the aspects of the installation, errors, often large errors, will certainly result. Finally, the installation of the controller (that is, screwing it to the wall) may not be performed by the same person who will be programming the unit. This throws further questions into the arena, and makes it difficult to ascertain how much time should be taken by each of the workers. If there is a cost overrun, whose fault is it?

The second solution to the high tech labor problem is much better, although it does involve one extra step in the estimating process. As you saw in the list of the elements of labor, there are ten basic functions that we bundle together into one labor unit. Our solution to the labor problem is to isolate and remove the variable element from the other nine elements of labor and charge for it separately.

By removing operation number 8 (testing) from our labor unit, and charging for it separately, we clean up the estimating process. All of our volatility is moved into one separate category, which demands thought before a labor unit is applied, and the rest of the labor, which is consistent and predictable, is covered by the basic labor unit.

As should be obvious, the title testing is hardly an accurate term to describe the types of operations our second labor figure is to account for. Testing would be only one component of these operations. Along with testing would be included any number of technical operations, such as the following:

Installing software

Programming

Configuring hardware

Configuring software

Training users

Running diagnostics

Transferring data

Creating reports

In addition to these, there could be any number of similar operations required. Over several years a general consensus has developed that this should be called "connect" labor.

Connect labor is where all of the volatility of high tech estimating goes. This labor should be calculated and charged separately from normal labor, which we are calling installation labor. Installation labor is the labor required to procure the material, move it to the installation area, tool up and mount the equipment in its place. In short, all of the normal elements of labor, except for the connect time.

Subcontracts

Technology installations frequently require the use of subcontractors. The reason for this is that there are so many new specialties, no one electrician or installer can be knowledgeable in all of them. Therefore, it is likely that all but the largest firms will encounter many situations where they have no one in their company who is knowledgeable about a certain type of application. In these circumstances, you will normally have to call the manufacturer of the system, who can refer you to a consultant or dealer of theirs in your local area, who can do this work for you.

The Takeoff

The process of taking off high tech systems is essentially the same as the process used for conventional estimating. By taking off, we mean the process of taking information off of a set of plans and/or specifications, and transferring it to estimate sheets. This requires the interpretation of graphic symbols on the plans, and transferring them into words and numbers that can be processed.

Briefly, the rules that apply to the takeoff process are as follows:

Review the symbol list. This is especially important for high tech work. High tech systems are not standardized and therefore vary widely. Make sure you know what the symbols you are looking at represent. This is fundamental.

Review the specifications. Obviously it is necessary to read a project's specifications, but it is also important to review the specifications before you begin your takeoff. Doing this may alert you to small details on the plans that you might otherwise overlook.

Mark all items that have been counted. Again, this is obvious, but a lot of people do this rather poorly. The object is to clearly and distinctly mark every item that has been counted. This must be done in such a way that you can instantly ascertain what has been counted. This means that you should color every counted item completely. Do not just put a check mark next to something you counted; color it in so fully that there will never be any room for question.

Always take off the most expensive items first. By taking the most expensive items off first, you are assuring that you will have numerous additional looks through the plans before you are done with them. Very often, you will find stray items that you missed on your first run through. You want as many chances as possible to find all of the costly items. This way, if you make a mistake, it will be less expensive.

Obtain quantities from other quantities whenever possible. For example, when you take off conduit, you don't try to count every strap that will be needed. Instead, you simply calculate how many feet of pipe will be required and then include one strap for every 7-10 feet of pipe. We call this obtaining a quantity from a quantity. Do it whenever you can. It will save you a good deal of time.

Do not rush. Cost estimating, by its very nature, is a slow, difficult process. In order to do a good estimate, you must do a careful, efficient takeoff.

Don't waste any time, but definitely don't go so fast that you miss things.

Maintain a good atmosphere. When performing estimates, it is very important to remain free of interruptions and to work in a good environment. Spending hours counting funny symbols on large, crowded sheets of paper is not particularly easy, so make it as easy on yourself as you can.

Develop mental pictures of the project. As you take off a project, picture yourself in the rooms, looking at the items you are taking off. Picture the item you are taking off in its place, its surroundings, the things around it, and how it connects to other items. If you get in the habit of doing this, you will greatly increase your skill.

Where to Charge Training

Many high tech installations require you to teach the owners of their representatives to use the system. This is difficult and particularly weighty upon high tech work, since you don't have to spend time teaching building owners how to use conventional electrical items, such as light switches, but you will certainly have to spend time teaching them how to use a sound system. Not only that, but you may have to supply operating instructions, teach a number of different people and answer numerous questions over the phone after the project is long over.

The question here is whether you include training charges in connect labor or include these costs as a separate job expense. This decision is essentially up to the discretion of the estimator. It is, however, usually best to charge general training to job expenses and to include incidental training in connect labor.

If you come from a background in the electrical construction industry, make sure that you accept these expenses as an integral part of your projects. Do not avoid them. The people who are buying your systems need this training, and they have a right to expect it. Include these costs in your estimates, and choose your most patient workers to do the training.

Charging Overhead

What percentage of overhead to assign to any type of electrical work (and how to assign it) can be a hotly-debated subject. Everyone seems to have their own opinion. Whatever percentage of overhead you charge, consider raising it a bit for high tech projects. As we have already said, the purchasing process is far more difficult for high tech work than it is for other, more established types of work. In addition to this, there are a number of other factors that are more difficult for high tech work than they are for more traditional types of work.

Almost every factor we can identify argues for including more overhead charges for high tech work. Not necessarily a lot more, but certainly something more. When we contract to do high tech installations, we are agreeing to go through uncharted, or at least partially uncharted, waters. This involves greater risk. And if we do encounter additional risk, it is only sensible to make sure we cover these risks. We do this by charging a little more for overhead and/or profit.

Labor Units

The labor units shown in this manual are necessarily average figures. They are based upon the following conditions:

1. **An average worker.**
2. **A maximum working height of twelve feet.**
3. **A normal availability of workers.**
4. **A reasonably accessible work area.**
5. **Proper tools and equipment.**
6. **A building not exceeding three stories.**
7. **Normal weather conditions.**

Any set of labor units must be tempered to the project to which they are applied. They are a starting point, not the final word. Difficult situations typically require an increase of 20-30%. Some very difficult installations may require even more. Especially good working conditions, or especially good workers may allow discounts to the labor units of 10-20%, and possibly more in some circumstances.

COMPUTER LABOR UNITS		
Item	**Install**	**Connect**
Computer (PC), processor only	1.00	1.0 – 3.0
Computer (PC), with monitor	1.40	1.2 – 3.2
Computer, minicomputer (AS400)	4.00	2.0 – 4.0
Computer, mainframe	12.00	N/A
Peripheral items		
replacement monitor and card	1.00	1.0 – 3.0
change video board	0.80	1.0 – 2.0
polarizing filter	0.40	N/A
add disk drive	2.00	1.0 – 2.0
add hard drive, with controller	2.50	1.5 – 4.5
add CD ROM drive	2.00	1.0 – 2.0
add bernoulli drive	2.00	1.0 – 2.0
add RAM, one card	2.00	1.4 – 3.0
add mouse, with software	0.70	.5 – 1.0
add trackball, with software	0.90	.5 – 1.0
change keyboard	0.40	.2 – .4
add internal modem	1.50	.5 – 2.0
upgrade CPU	1.50	1.0 – 3.0
add expansion board	1.40	.5 – 1.5
desktop scanner	1.00	.5 – 1.0
expansion chassis	1.40	.5 – 1.5

COMPUTER LABOR UNITS *(cont.)*		
Item	**Install**	**Connect**
internal tape backup	1.50	1.0 – 2.0
external tape backup	0.70	.7 – 1.8
replace power supply	2.00	N/A
bar code reader, hand-held	1.00	.4 – 2.0
sound board	1.00	.7 – 2.0
remote speaker	0.50	N/A
pen tablet	0.50	.5 – 1.4
Miscellaneous		
install software	N/A	1.0 – 4.0+
modular cables	0.15	N/A
cable adapters	0.35	N/A
clean heads, floppy drive	0.70	0.10
clean heads, hard drive	1.70	0.30
Network devices		
network interface card	1.00	.5 – 1.0
host adapter, with cable	1.00	1.0 – 2.4
PAL kit	2.00	2.0 – 4.0
server disk kit	3.00	2.0 – 4.0
lap link	1.00	.7 – 1.4
LAN hub concentrator	2.00	1.9 – 4.5
multiplexer	2.40	1.9 – 4.5

COMPUTER LABOR UNITS *(cont.)*		
Item	**Install**	**Connect**
telecluster	3.00	2.0 – 5.0
ethernet adapter	0.50	N/A
wireless transceiver	1.40	1.4 – 2.9
long distance adapter	1.00	.5 – 1.0
signal booster	1.00	.5 – 1.0
line splitter	1.40	1.0 – 4.0
bridge	2.00	2.0 – 7.0
router	2.00	2.0 – 7.0
WAN interface card	1.20	2.0 – 5.0
repeater	1.70	2.0 – 5.0
diagnostic modem	1.30	.9 – 3.0
Printers		
laser printer	1.10	.5 – 2.0
color printer	1.10	.8 – 4.0
A/B printer switch, one cable	1.10	.6 – 1.5
multi-printer box	1.40	.6 – 1.6
pen plotter	1.50	1.0 – 3.5
printer controller	1.00	.9 – 2.0
Network cables, per foot, accessible locations		
unshielded twisted pair	0.01	N/A
shielded twisted pair	0.01	N/A

COMPUTER LABOR UNITS *(cont.)*		
Item	**Install**	**Connect**
coaxial	0.012	N/A
4-conductor telephone cable	0.01	N/A
other types	0.012	N/A
Miscellaneous, cable related		
coaxial crimp connectors	0.08	N/A
coaxial twist-on connectors	0.08	N/A
coaxial tee connectors	0.10	N/A
twinaxial connectors	0.15	N/A
shielded twisted pair termination	0.12	N/A
unshielded twisted pair termination	0.10	N/A
punchdown block, no termination	1.00	N/A
wallplate with jack	0.80	N/A
modular connectors, telephone	0.14	N/A
in-line couplers	0.07	N/A
male/female adapters	0.40	N/A
cable end assemblies	0.65	N/A
extension	0.20	N/A

COMPUTER ROOMS		
Item	**Normal**	**Difficult**
2' x 2' raised flooring		
2' vertical support	0.50	0.63
perimeter channel, per lineal foot	0.08	0.10
mounting track, per lineal foot	0.05	0.08
2 x 2 squares	0.20	0.25
Isolated ground receptacles		
20 amp duplex	0.20	0.25
20 amp, 3 wire	0.20	0.25
20 amp, 4 wire	0.24	0.30
20 amp, 5 wire	0.30	0.38
30 amp, 3 wire	0.24	0.30
30 amp, 4 wire	0.30	0.38
30 amp, 5 wire	0.35	0.44
40 amp, 3 wire	0.30	0.38
40 amp, 4 wire	0.35	0.44
40 amp, 5 wire	0.40	0.50
50 amp, 3 wire	0.35	0.44
50 amp, 4 wire	0.40	0.50
50 amp, 5 wire	0.45	0.56

COMPUTER ROOMS *(cont.)*		
Item	**Normal**	**Difficult**
Cord, per foot		
#12/3 (SJ, SJO, SO, etc.)	0.1	0.012
12/4	0.012	0.015
12/5	0.2	0.25
10/3	0.012	0.015
10/4	0.015	0.019
10/5	0.024	0.03
8/3	0.02	0.025
8/4	0.025	0.031
8/5	0.3	0.038

COMPUTER GROUNDING		
Item	Normal	Difficult
Ground rods		
½" x 8'	0.60	0.75
½" x 10'	0.70	0.88
⅝" x 8'	0.65	0.81
⅝" x 10'	0.75	0.94
¾" x 8'	0.70	0.88
¾" x 10'	0.85	1.06
½" ground rod clamps	0.15	0.19
⅝" ground rod clamps	0.18	0.23
¾" ground rod clamp	0.20	0.25
1" – 3" water pipe clamp	0.50	0.63
4" – 6" water pipe clamp	0.70	0.88
Grounding hubs		
¾"	0.20	0.25
1"	0.20	0.25
1¼"	0.25	0.32
1½"	0.30	0.38
2"	0.35	0.44

COMPUTER GROUNDING *(cont.)*		
Item	**Normal**	**Difficult**
Grounding bushings		
½"	0.10	0.13
¾"	0.12	0.15
1"	0.15	0.19
1¼"	0.18	0.23
1½"	0.21	0.26
2"	0.24	0.30
2½"	0.27	0.34
3"	0.30	0.38
3½"	0.35	0.44
4"	0.40	0.50
Miscellaneous		
ground clips	0.02	0.025
ground screws	0.04	0.05
grounding pigtail	0.04	0.05
core drilling for ground rod	1.00	1.25
grounding receptacle, in floor	1.50	1.88
grounding receptacle, wall	1.00	1.25

COMPUTER POWER		
Item	Normal	Difficult
Metal oxide varistors		
0 – 100 volts (DC or AC rms)	0.16	0.20
over 100 volts (DC or AC rms)	0.20	0.25
Zener diodes		
0 – 5 watts	0.16	0.20
over 5 watts	0.20	0.25
Solderless terminals (sta-kons)		
#22 – 18	0.05	0.06
#16 – 14	0.05	0.06
#12 – 10	0.05	0.06
Lightning arresters, panel mount		
120/240 volt, 1 phase, 3 wire	0.80	1.00
120/208 volt, 3 phase, 4 wire	1.00	1.25
120/240 volt, 3 phase, 4 wire	1.00	1.25
277/480 volt, 3 phase, 4 wire	1.10	1.38
480 volt, 3 phase, 3 wire	1.00	1.25
600 volt, 3 phase, 3 wire	1.20	1.50
Spike suppressor, panel mount		
120/240 volt, 1 phase, 3 wire	0.80	1.00
120/208 volt, 3 phase, 4 wire	1.00	1.25
120/240 volt, 3 phase, 4 wire	1.00	1.25

COMPUTER POWER *(cont.)*		
Item	**Normal**	**Difficult**
277/480 volt, 3 phase, 4 wire	1.10	1.38
480 volt, 3 phase, 3 wire	1.00	1.25
600 volt, 3 phase, 3 wire	1.20	1.50
Cable protector	0.40	0.50
Low voltage suppressor module	0.30	0.38
Surge suppression receptacle	0.24	0.30
Outlet strip with suppressors	0.10	0.13
Uninterruptible power sources		
mini-UPS	2.50	3.10
3 KVA	9.00	11.25
5 KVA	16.00	20.00
7.5 KVA	22.00	25.50
10 KVA	28.00	35.00
15 KVA	42.00	50.50
25 KVA	70.00	87.50
Rotary UPS systems		
7.5 KVA	20.00	25.00
10 KVA	24.00	30.00
15 KVA	35.00	43.75
25 KVA	50.00	62.50

COMPUTER POWER *(cont.)*		
Item	**Normal**	**Difficult**
Power centers, ferroresonant		
5 KVA	10.00	12.50
7.5 KVA	12.00	15.00
10 KVA	16.00	20.00
15 KVA	20.00	25.00
25 KVA	25.00	31.25
Shielded isolation transformers		
10 KVA	5.00	6.25
15 KVA	7.00	8.75
25 KVA	9.00	11.25
30 KVA	11.00	13.75
45 KVA	14.00	17.50
75 KVA	18.00	22.50
112.5 KVA	22.00	27.50
150 KVA	25.00	31.25
225 KVA	30.00	37.50
300 KVA	35.00	43.75
350 KVA	40.00	50.00

COMPUTER POWER *(cont.)*		
Item	**Normal**	**Difficult**
Transformer isolation isolators		
15 KVA	0.70	0.88
30 KVA	0.80	1.00
45 KVA	0.90	1.13
75 KVA	1.00	1.25
112.5 KVA	1.10	1.38
150 KVA	1.20	1.50
225 KVA	1.30	1.63
300 KVA	1.40	1.75
350 KVA	1.50	1.88
Voltage regulators		
500 VA	5.00	6.25
1000 VA	7.00	8.75
2000 VA	8.00	10.00
5000 VA	12.00	15.00

ELECTRONICS LABOR UNITS		
Item	**Open**	**Enclosed**
Plug-in relay	0.10	0.25
Relay socket	0.18	0.30
Plug-in devices		
audio amp	0.10	0.25
pre-amp	0.10	0.25
oscillator	0.10	0.25
equalizer	0.10	0.25
Power relays		
20 amp, 2 pole	0.40	0.55
20 amp, 4 pole	0.50	0.65
30 amp, 8 pole	1.00	1.15
Rechargeable batteries	0.10	0.25
Transformers		
0 – 100 VA	0.30	0.45
150 – 500 VA	0.50	0.65
1000 VA	1.00	1.25
1500 VA	1.25	1.45
Resistors		
0 – 2 watts	0.10	N/A
3 – 5 watts	0.12	N/A

ELECTRONICS LABOR UNITS *(cont.)*

Item	Open	Enclosed
Rheostats and potentiometers		
0 – 5 watts	0.20	N/A
6 – 10 watts	0.21	N/A
11 – 25 watts	0.30	0.50
Electrolytic capacitors		
0 – 500 mfd	0.16	0.40
over 500 mfd	0.25	0.50
Other capacitors		
0 – 2 mfd	0.10	0.25
Ribbon cable assemblies, 25-pin	0.24	0.44
Cooling fans, 3 – 4"	0.40	0.60
DIP plugs		
14 – 16 pin, with terminations	0.40	0.50
24 pin, terms	0.50	0.60
40 pin, terms	0.60	0.70
Edge cards		
10 contacts, terms	0.40	0.50
20 contacts, terms	0.50	0.60
40 contacts, terms	0.65	0.75
60 contacts, terms	0.75	0.85

ELECTRONICS LABOR UNITS *(cont.)*		
Item	**Open**	**Enclosed**
Socket connectors		
10 contacts, terms	0.40	0.50
20 contacts, terms	0.50	0.60
40 contacts, terms	0.65	0.75
60 contacts, terms	0.75	0.85
Latch headers		
10 contacts, terms	0.40	0.50
20 contacts, terms	0.50	0.60
40 contacts, terms	0.65	0.75
60 contacts, terms	0.75	0.85
Panel mount plug (AB)	0.20	0.30
Panel mount socket (AB)	0.20	0.30
Cable mount plug (CCT)	0.22	0.32
Cable mount socket (CCT)	0.22	0.32
Panel mount AC receptacles	0.20	0.30
Diodes		
0 – 1 amp average current	0.10	N/A
1 – 3 amp average current	0.12	N/A
Silicon controller rectifiers		
0 – 5 amp average current	0.12	0.24
6 – 10 amp average current	0.17	0.29

ELECTRONICS LABOR UNITS *(cont.)*		
Item	**Open**	**Enclosed**
11 – 25 amp average current	0.24	0.36
26 – 40 amp average current	0.40	0.50
Metal oxide varistors		
0 – 100 volts (DC or AC rms)	0.12	N/A
over 100 volts (DC or AC rms)	0.14	N/A
Zener diodes, 0 – 5 watts	0.12	N/A
Thermal fuses, 15 amp max.	0.12	N/A
Infrared diodes	0.12	N/A
Phototransistor detectors	0.14	N/A
Optoisolators	0.14	N/A
LEDs	0.11	N/A
Plasma display, 9 digits	0.24	0.34
Mini lamps	0.14	N/A
Pilot lights	0.14	N/A
Silicon unijunction transistor		
50 ma emitter current	0.14	0.25
Transistors		
max. 4 collector amps	0.14	0.25
max. 10 collector amps	0.18	0.29
max. 20 collector amps	0.20	0.33

ELECTRONICS LABOR UNITS *(cont.)*		
Item	**Open**	**Enclosed**
IC chips		
8 pin	0.38	0.53
14 pin	0.40	0.55
20 pin	0.50	0.65
40 pin	0.60	0.75
IC sockets		
8 pin	0.20	0.35
14 pin	0.25	0.40
20 pin	0.30	0.45
40 pin	0.35	0.50
Voltage regulators		
2 pin	0.15	0.30
8 pin	0.24	0.40
14 pin	0.30	0.45
Power supplies		
50 watt	0.40	0.55
100 watt	0.50	0.65
200 watt	0.60	0.75
Speakers		
18 watt	0.20	0.34
30 watt	0.25	0.39

ELECTRONICS LABOR UNITS *(cont.)*		
Item	**Open**	**Enclosed**
50 watt	0.30	0.45
80 watt	0.35	0.50
Toggle switches		
SPST	0.15	0.25
DPDT	0.17	0.27
3PDT	0.19	0.29
4PDT	0.21	0.31
Snap switches		
SPST	0.15	0.25
SPDT	0.15	0.25
Solderless terminals		
#22 – 18	0.05	N/A
#16 – 14	0.05	N/A
#12 – 10	0.05	N/A
Terminal blocks, barrier type		
2 circuit	0.14	0.24
4 circuit	0.16	0.26
8 circuit	0.18	0.28
12 circuit	0.20	0.30
20 circuit	0.24	0.34

ELECTRONICS LABOR UNITS *(cont.)*		
Item	**Open**	**Enclosed**
Panel meters, 2½" square	0.40	0.50
Standard phone jack	0.15	N/A
Shielded phone jack	0.17	N/A
3 conductor phone jack	0.17	N/A
In-line jack	0.17	N/A
2 conductor chassis jack	0.18	N/A
3 conductor chassis jack	0.20	N/A
Wall plate		
single device	0.18	N/A
two devices	0.24	N/A
Fuseblock, HTB type	0.18	0.27
In-line fuseholder, GMQ or GLR	0.14	0.23
Glass tube fuses (AGC, MSL, MDL, MDA)		
0 – 30 amp	0.03	0.15
Axial lead fuses, 0 – 4 amp	0.10	0.22
Cartridge fuses		
0 – 30 amp	0.05	0.15
31 – 60 amp	0.05	0.15
61 – 100 amp	0.07	0.15
101 – 200 amp	0.07	0.15

ELECTRONICS LABOR UNITS *(cont.)*		
Item	**Open**	**Enclosed**
Gromets	0.05	N/A
Bumpers	0.04	N/A
Metal chassis boxes		
4" x 2"	0.30	0.40
5" x 10"	0.40	0.50
5" x 17"	0.50	0.60
Heat shrink tubing		
⅛" x 6" long	0.15	0.25
½" x 6" long	0.17	0.27
Spiral wrap, 3' length	0.14	0.24
Hook-up wire, per piece	0.05	0.07
Heat sink, 1½" square	0.10	0.15
Miscellaneous		
misc. A/V cables	0.20	N/A
ribbon cable	0.24	N/A
in-line couplers	0.07	N/A
cable adaptors, various types	0.25	N/A

FIBER OPTICS		
Item	**Normal**	**Difficult**
Optical fiber cables, per foot		
1 – 4 fibers, in conduit	0.016	0.02
1 – 4 fibers, accessible locations	0.014	0.018
12 – 24 fibers, in conduit	0.02	0.025
12 – 24 fibers, accessible locations	0.018	0.023
48 fibers, in conduit	0.03	0.038
48 fibers, accessible locations	0.025	0.031
72 fibers, in conduit	0.04	0.05
72 fibers, accessible locations	0.032	0.04
144 fibers, in conduit	0.05	0.065
144 fibers, accessible locations	0.04	0.05
Hybrid cables		
1 – 4 fibers, in conduit	0.02	0.025
1 – 4 fibers, accessible locations	0.017	0.021
12 – 24 fibers, in conduit	0.024	0.03
12 – 24 fibers, accessible locations	0.022	0.028
Testing, per fiber	0.20	0.30

FIBER OPTICS *(cont.)*

Item	Normal	Difficult
Splices, inc. prep and failures, trained workers		
fusion	0.40	0.50
mechanical	0.50	0.65
array splice, 12 fibers	1.30	1.65
Coupler (connector-connector)	0.20	0.30
Terminations, inc. prep and failures, trained workers		
polishing required	0.60	0.80
no-polish connectors	0.50	0.65
FDDI dual connector	1.00	1.25
Miscellaneous		
cross-connect box, 144 fibers	3.00	4.00
splice cabinet	2.00	2.50
splice case	1.80	2.25
breakout kit, 6 fiber	1.00	1.40
tie-wraps	0.01	0.02
wire markers	0.01	0.01

TELECOM		
Item	Install	Connect
Telephones		
single line desk phone	0.20	N/A
2 line desk phone	0.24	N/A
4 – 8 line desk phone	0.30	N/A
central console	1.20	.4 – 1.4
wall mount kit	0.20	N/A
cordless phone	0.24	.1 – .3
replace handset	0.10	N/A
noisy environment handset	0.10	.1 – .2
headset	0.40	.1 – .2
speaker phone module	0.30	.1 – .2
music-on-hold unit	0.30	.2 – .5
conference speaker/mic unit	0.30	.1 – .3
conference phone	0.20	.2 – .4
table conference unit	0.40	.2 – .4
answering machine	0.30	.1 – .2
small paging system	0.90	.8 – 2.0
speaker phone add-on kit	1.20	.1 – .5
Large systems		
auto switching unit	1.40	2.0 – 8.0
remote control for above	0.50	1.0 – 3.0

TELECOM *(cont.)*		
Item	**Install**	**Connect**
8 line central unit	2.00	4.0 – 12.0
16 line central unit	3.00	5.0 – 14.0
24 line central unit	4.00	7.0 – 16.0
40 line central unit	6.00	10.0 – 20.0
80 line central unit	10.00	16.0 – 28.0
digital line card	0.50	.1 – .4
trunk circuit card	0.50	.1 – .3
other cards	0.50	.1 – .4
call accounting software	2.00	2.0 – 8.0
function modules	0.80	.1 – .4
line modules	0.80	.1 – .3
paging system	1.00	1.0 – 2.5
Fax machine	0.60	.2 – .5
Cables, accessible locations		
unshielded twisted pair	0.01	N/A
shielded twisted pair	0.01	N/A
coaxial	0.012	N/A
4-conductor telephone cable	0.01	N/A
other types	0.012	N/A
Miscellaneous, cable related		
shielded twisted pair termination	0.12	N/A

TELECOM *(cont.)*		
Item	**Install**	**Connect**
unshielded twisted pair termination	0.10	N/A
punchdown block, no termination	1.00	N/A
wallplate with jack	0.80	N/A
modular connectors, telephone	0.14	N/A
in-line couplers	0.07	N/A
male/female adapters	0.40	N/A
cable end assemblies	0.65	N/A
extension	0.20	N/A
punchdown block, no termination	1.00	N/A
dual data connection block	0.30	N/A
flush connecting block	0.25	N/A
2 line flush connecting block	0.30	N/A
modular telephone connectors	0.14	N/A
phone privacy device	1.00	N/A
powerline carrier phone	0.80	N/A
telephone-based door control	2.40	1.0 – 2.8
500 VA UPS system	2.00	.7 – 2.0
19" rack, 2'	1.80	N/A
19" rack, 4'	3.00	N/A
19" rack, 6'	4.50	N/A
tie-wraps	0.01	N/A
wire markers	0.01	N/A

TV SYSTEMS		
Item	**Install**	**Connect**
Camera equipment		
video camera	0.60	.2 – .8
power bracket	0.50	N/A
camera control unit	0.60	.3 – 1.0
standard mounting	0.60	N/A
swivel camera mount	0.70	N/A
tamper-proof mounting	1.50	N/A
automatic pan mounting	1.00	.2 – .8
add or replace lens	0.40	N/A
auto-iris feature	N/A	.2 – .4
Monitors		
9" monitor	0.60	.1 – .3
12" monitor	0.70	.1 – .3
15" monitor	0.80	.1 – .3
19" monitor	0.90	.1 – .3
monitor mounting bracket	1.00	N/A
Switches and sequencers		
4 camera switch, manual	1.00	N/A
8 camera switch, manual	1.40	N/A
4 camera sequencer	1.00	.2 – .4
8 camera sequencer	1.40	.3 – .6

TV SYSTEMS *(cont.)*		
Item	**Install**	**Connect**
synchronizing feature	N/A	.2 – .4
bridging feature	N/A	.1 – .3
Videocassette recorder	1.00	.3 – .7
Miscellaneous		
TV receptacle	0.30	N/A
2 wire plug or connector	0.20	N/A
4 wire plug or connector	0.35	N/A
amplifier	0.70	N/A
coupler	0.70	N/A
splitter	0.80	N/A
coax surge suppressor	0.40	N/A
video motion detector	0.80	.2 – .4
screen splitter	0.50	.1 – .3
head end equipment	4.00	2.0 – 4.0
antenna	6.00	1.0 – 4.0
console	3.00	3.0 – 7.0
satellite receiver, 4'	8.00	2.0 – 4.0
satellite receiver, 6'	10.00	2.0 – 4.0
satellite receiver, 8'	12.00	2.0 – 4.0
19" rack adaptors	0.50	N/A
19" rack, 2'	1.80	N/A

TV SYSTEMS *(cont.)*		
Item	**Install**	**Connect**
19" rack, 4'	3.00	N/A
19" rack, 6'	4.50	N/A
tie-wraps	0.01	N/A
wire markers	0.01	N/A
Cables, per foot, accessible locations		
coaxial	0.012	N/A
4-conductor telephone cable	0.01	N/A
other types	0.01	N/A
#14 single conductor in conduit	0.01	N/A
#16 single conductor in conduit	0.01	N/A
#18 single conductor in conduit	0.01	N/A
Miscellaneous, cable related		
shielded twisted pair termination	0.12	N/A
unshielded twisted pair termination	0.10	N/A
modular connectors, telephone	0.14	N/A
in-line couplers	0.07	N/A
male/female adapters	0.40	N/A
coaxial crimp connectors	0.08	N/A
coaxial twist-on connectors	0.08	N/A
coaxial tee connectors	0.10	N/A
twinaxial connectors	0.15	N/A

SOUND SYSTEMS		
Item	**Install**	**Connect**
Signal generators		
cassette deck, single or double	1.00	N/A
CD player	1.00	N/A
radio receiver	1.00	N/A
microphone	0.40	N/A
wireless microphone	0.40	N/A
wireless mike receiver	1.00	.1 – .5
reel-to-reel tape player	1.20	N/A
audio cart player	1.00	N/A
power supply	1.00	N/A
remote controller	2.00	1.4 – 4.0
Signal processing equipment		
power amp	2.40	N/A
pre-amp	1.80	N/A
equalizer	1.00	N/A
4 channel mixer	1.00	N/A
8 channel mixer	1.50	N/A
12 channel mixer	2.00	N/A
16 channel mixer	2.50	N/A
24 channel mixer	4.00	N/A
volume control	0.30	N/A

SOUND SYSTEMS *(cont.)*		
Item	**Install**	**Connect**
programmable switch	0.70	.4 – .8
speaker selection switch	0.80	N/A
noise reduction unit	2.00	.2 – .5
equalizer, 1 channel	1.00	N/A
equalizer, 2 channel	1.10	N/A
equalizer, 4 channel	1.30	N/A
filters, various types	0.80	N/A
delay unit	1.00	.2 – 1.0
aural exiter	1.00	N/A
crossover	1.00	N/A
VU meters	0.80	N/A
synch generator	1.00	N/A
computer interface	2.00	2.0 – 8.0
routing system	1.70	1.0 – 4.0
Sound generating equipment		
ceiling speaker	0.60	N/A
wall speaker	0.40	N/A
paging speaker	0.40	N/A
column speaker	3.50	N/A
horn	0.40	N/A
double horn	0.50	N/A

SOUND SYSTEMS *(cont.)*

Item	Install	Connect
Cables, per foot, accessible locations		
unshielded twisted pair	0.01	N/A
shielded twisted pair	0.01	N/A
coaxial	0.01	N/A
4-conductor telephone cable	0.01	N/A
other types	0.01	N/A
#14 single conductor in conduit	0.01	N/A
#16 single conductor in conduit	0.01	N/A
#18 single conductor in conduit	0.01	N/A
Miscellaneous, cable related		
shielded twisted pair termination	0.12	N/A
unshielded twisted pair termination	0.10	N/A
modular connectors, telephone	0.14	N/A
in-line couplers	0.07	N/A
male/female adapters	0.40	N/A
microphone outlet, flush	0.80	N/A
intercom, central unit	4.00	4.0 – 8.5
intercom, satellite unit	0.70	N/A
wireless speaker system	2.00	.8 – 1.8
antenna	6.00	1.0 – 4.0
19" rack adaptors	0.50	N/A

SOUND SYSTEMS *(cont.)*		
Item	**Install**	**Connect**
microphone stand	0.20	N/A
plywood backboard	1.50	N/A
19" rack, 2'	1.80	N/A
19" rack, 4'	3.00	N/A
19" rack, 6'	4.50	N/A
tie-wraps	0.01	N/A
wire markers	0.01	N/A

SIGNALING SYSTEMS		
Item	**Install**	**Connect**
Control panel, general	2.00	1.0 – 4.0
Power amp	2.40	N/A
Pre-amp	1.80	N/A
Equalizer	1.00	N/A
4-channel mixer	1.00	N/A
Volume control	0.30	N/A
Programmable switch	0.70	.4 – .8
Speaker selection switch	0.80	N/A
Filters, various types	0.80	N/A
VU meters	0.80	N/A
Computer interface	2.00	2.0 – 8.0
Ceiling speaker	0.60	N/A
Wall speaker	0.40	N/A
Paging speaker	0.40	N/A
Horn	0.40	N/A
Bell	0.40	N/A
Buzzer	0.35	N/A
Push button	0.30	N/A
WP push button	0.34	N/A
Mortise door opener	0.40	N/A
Transformer, 25 VA	0.30	N/A

SIGNALING SYSTEMS *(cont.)*		
Item	**Install**	**Connect**
Clock systems		
single-face clock	0.60	N/A
double-face clock	0.60	N/A
skeleton clock	2.00	N/A
master clock	3.50	N/A
signal generator	2.40	.4 – .9
elapsed time indicator	0.60	N/A
elapsed time control	0.40	N/A
clock and speaker combo	0.80	N/A
flasher	0.40	N/A
control board	2.00	.2 – .7
program unit	3.00	.2 – .8
clock back box	0.20	N/A
wire guard	0.15	N/A
Nurse call system		
single bed station	0.40	N/A
dual bed station	0.50	N/A
call-in cord	0.15	N/A
pull cord	0.15	N/A
pillow speaker	0.15	N/A
dome light	0.40	N/A

SIGNALING SYSTEMS *(cont.)*		
Item	**Install**	**Connect**
zone light	0.40	N/A
staff station	0.40	N/A
duty station	0.40	N/A
utility station	0.40	N/A
nurse's station	0.50	N/A
surgical station	0.50	N/A
master station	1.80	N/A
control station	0.60	N/A
annunciator	1.50	N/A
power supply	1.00	N/A
speakers	0.40	N/A
foot switch	0.20	N/A
Code blue systems		
bed station	0.50	N/A
dome light	0.40	N/A
zone light	0.40	N/A
pull cord	0.15	N/A
nurse's station	0.40	N/A
annunciator	1.50	N/A
power supply	1.00	N/A

SIGNALING SYSTEMS *(cont.)*		
Item	**Install**	**Connect**
Hospital ground detection		
power ground modules	1.60	N/A
slave ground module	1.20	N/A
master ground module	1.00	N/A
remote indicator	1.00	N/A
X-ray indicator	1.40	N/A
micro-ammeter	1.40	N/A
supervisory module	1.00	N/A
ground cords	0.14	N/A
Hospital isolation monitors		
5 ma, 120 volts	2.40	N/A
5 ma, 208 volts	2.40	N/A
5 ma, 240 volts	2.40	N/A
Nurse's indicator alarm annunciators		
4 ckt. flush	2.40	N/A
6 ckt. flush	4.00	N/A
12 ckt. flush	5.80	N/A
4 ckt. desktop	2.00	N/A
6 ckt. desktop	2.40	N/A
12 ckt. desktop	3.60	N/A

SIGNALING SYSTEMS *(cont.)*		
Item	**Install**	**Connect**
Digital clocks and timers		
separate display	1.20	N/A
one display	1.20	N/A
remote control	0.80	N/A
battery pack	0.80	N/A
Surgical chronometer		
clock and 3 timers	1.80	N/A
auxiliary control	0.80	N/A
Cables, per foot, accessible locations		
unshielded twisted pair	0.01	N/A
shielded twisted pair	0.01	N/A
coaxial	0.012	N/A
4-conductor telephone cable	0.01	N/A
other types	0.01	N/A
#14 single conductor in conduit	0.005	N/A
#16 single conductor in conduit	0.005	N/A
#18 single conductor in conduit	0.005	N/A

SIGNALING SYSTEMS *(cont.)*

Item	Install	Connect
Miscellaneous, cable related		
shielded twisted pair termination	0.12	N/A
unshielded twisted pair termination	0.10	N/A
modular connectors, telephone	0.14	N/A
in-line couplers	0.07	N/A
male/female adapters	0.40	N/A
microphone outlet, flush	0.80	N/A
intercom, central unit	4.00	4.0 – 8.5
intercom, satellite unit	0.70	N/A
19" rack adaptors	0.50	N/A
plywood backboard	1.50	N/A
19" rack, 2'	1.80	N/A
19" rack, 4'	3.00	N/A
19" rack, 6'	4.50	N/A
tie-wraps	0.01	N/A
wire markers	0.01	N/A

FIRE ALARMS		
Item	**Install**	**Connect**
Controllers		
4 zone	1.25	1.0 – 4.0
8 zone	1.50	2.0 – 6.0
12 zone	2.00	3.0 – 8.0
24 zone	3.00	5.0 – 10.0
48 zone	4.00	8.0 – 12.0
power supply	1.00	N/A
battery	1.00	.4 – .8
battery charger	0.80	.4 – .8
relay	0.40	N/A
test switch	0.30	.1 – .2
remote indicator	0.30	.1 – .2
fireman's phone	1.20	.4 – .6
add digital communicator	0.80	.1 – .4
remote interface	0.40	.2 – .7
function circuit card	0.50	.2 – .5

FIRE ALARMS *(cont.)*		
Item	**Install**	**Connect**
Annunciators		
8 zone	1.00	1.0 – 3.0
12 zone	1.30	1.5 – 4.0
16 zone	1.80	2.0 – 4.0
24 zone	2.00	2.5 – 4.5
48 zone	2.50	3.0 – 5.0
Detection devices		
manual pull station	0.35	N/A
explosionproof pull station	0.80	N/A
smoke detector	0.40	N/A
remote LED detector	0.30	N/A
heat detector	0.40	N/A
thermal detector	0.40	N/A
ionization detector	0.40	N/A
duct detector	2.50	N/A
remote test station	0.50	.2 – .4

FIRE ALARMS *(cont.)*		
Item	**Install**	**Connect**
Signaling devices		
horn	0.50	N/A
weatherproof horn	0.60	N/A
amplified horn	0.80	N/A
bell	0.50	N/A
WP bell	0.60	N/A
siren	0.80	N/A
chime	0.50	N/A
audio/visual	0.50	N/A
strobe light	0.50	N/A
explosionproof strobe	0.70	N/A
rotating beacon	0.70	N/A
Miscellaneous		
door holder	0.50	N/A
speaker	0.50	N/A
flow switch, no plumbing	0.50	N/A
tamper switch	1.00	N/A
adapter plate	0.20	N/A
trim plate	0.20	N/A
single horn projector	0.25	N/A
dual horn projector	0.30	N/A

FIRE ALARMS *(cont.)*		
Item	**Install**	**Connect**
Cables, per foot, accessible locations		
18/2 fire cable	0.008	N/A
18/4 fire cable	0.008	N/A
18/6 fire cable	0.009	N/A
18/8 fire cable	0.01	N/A
16/2 fire cable	0.008	N/A
16/4 fire cable	0.01	N/A
other cables	0.01	N/A
#14 single conductor in conduit	0.005	N/A
#16 single conductor in conduit	0.005	N/A
#18 single conductor in conduit	0.005	N/A
Miscellaneous, cable related		
shielded twisted pair termination	0.12	N/A
unshielded twisted pair termination	0.10	N/A
modular connectors, telephone	0.14	N/A
in-line couplers	0.07	N/A
male/female adapters	0.40	N/A

SECURITY SYSTEMS

Item	Install	Connect
Control panels		
residential/small commercial	2.00	1.0 – 2.0
medium commercial	2.50	1.5 – 3.0
large commercial/institutional	3.00	2.0 – 4.0
separate annunciator panel	1.75	1.0 – 2.0
remote controller unit	1.75	2.0 – 4.0
residential wireless console	1.00	1.4 – 2.9
residential wireless panel	1.75	1.5 – 3.4
commercial wireless panel	2.00	2.0 – 3.8
change battery backup	1.00	.4 – .8
replace circuit board	0.70	.4 – .8
add dialer card	1.00	.7 – 1.4
zone expansion module	1.00	.8 – 1.6
logic module	0.90	.9 – 1.9
AC failure module	0.80	.4 – 1.0
timer monitor module	0.80	.4 – .8
single line phone monitor	0.80	.3 – .8
2 line phone monitor	0.90	.4 – .9
ground start module	0.80	.4 – .8
fire supervision module	0.80	.4 – .8
8 zone phone line monitor	0.80	.7 – 1.7

SECURITY SYSTEMS *(cont.)*		
Item	**Install**	**Connect**
status beeper	0.20	.1 – .2
electronic timer	0.50	.2 – .5
Peripherals		
separate digital communicator	1.40	.7 – 2.0
cellular phone link	3.00	2.0 – 4.0
telephone control module	0.90	1.0 – 3.0
single channel receiver	0.50	.2 – .5
2 channel receiver	0.60	.4 – .6
4 channel receiver	0.70	.6 – .9
8 channel receiver	0.80	.8 – 1.2
antenna pre-amp	0.70	.5 – 1.2
midrange (5 mile) transmitter	1.20	.5 – 1.2
midrange receiver	1.20	.5 – 1.2
extended range antenna	0.70	.2 – .4
lightning protection module	0.80	N/A
Detection devices		
magnetic switch	0.20	N/A
wireless magnetic switch	0.30	.1 – .2
glass break detector	0.20	N/A
wireless glass break detector	0.30	.1 – .2
seismic detector	0.30	N/A
smoke detector	0.50	N/A

SECURITY SYSTEMS *(cont.)*		
Item	**Install**	**Connect**
wireless smoke detector	0.60	.1 – .2
heat sensor	0.50	N/A
freeze sensor	0.50	N/A
water sensor	0.60	N/A
keypad	0.40	N/A
wireless keypad	0.50	.1 – .2
infrared motion sensor	0.50	N/A
wireless infrared motion sensor	0.60	.1 – .2
microwave/PIR detector	0.50	N/A
driveway sensor	3.00	1.0 – 2.0
floormat pressure sensor	0.40	N/A
sound sensor	0.40	N/A
handheld transmitter	0.20	.15 – .3
cash drawer transmitter	0.20	.15 – .3
wireless money trap	0.20	.15 – .3
wall-mounted transmitter	0.40	.1 – .2
cabinet tamper switch	0.30	N/A
valve tamper switch	2.40	N/A
holdup button	0.80	N/A
remote switch	0.40	N/A
push button, wall mount	0.50	N/A
door touchbar	1.00	N/A

SECURITY SYSTEMS *(cont.)*		
Item	**Install**	**Connect**
Signaling devices		
weatherproof outdoor horn	0.70	N/A
indoor horn	0.40	N/A
wireless siren	0.45	.1 – .2
indoor strobe	0.50	N/A
X-10 burglar alarm module	0.70	.2 – 1.0
X-10 switch modules, new work	0.25	.1 – .2
X-10 switch modules, rehab	0.40	.1 – .3
Access control devices		
card reader, wall mount	0.70	.2 – 1.0
card reader, post mount	0.90	.2 – 1.0
reader controller	1.00	.4 – 1.4
palmprint identifier	2.00	2.0 – 5.0
program system computer	N/A	4.0 – 10.0
telephone door security unit	4.00	5.0 – 10.0
remote intercom unit	2.00	.5 – 1.4
Miscellaneous		
door lock protector	1.00	N/A
key switch	1.10	N/A
door closer	2.00	N/A
electric door strike, new work	1.20	N/A

SECURITY SYSTEMS *(cont.)*		
Item	**Install**	**Connect**
electric door strike, refit	2.20	N/A
electric mortice lock, new	2.00	N/A
electric mortice lock, refit	3.00	N/A
end-of-line resistors	0.10	N/A
X-10 Entry Guard receiver	0.15	N/A
X-10 Entry Guard transmitter	0.40	.1 – .4
scrambling telephone	0.50	N/A
electronic eye, 2-piece	1.20	N/A
touchpad door lock, new work	1.40	N/A
touchpad door lock, refit	2.40	N/A
card doorlock, new work	1.40	N/A
card doorlock, refit	2.40	N/A
electric gate, controls only	2.50	1.0 – 3.0
curbside keypad, no concrete	2.00	N/A
curbside card reader, no concrete	2.00	N/A
Cables, per foot, accessible locations		
unshielded twisted pair	0.01	N/A
shielded twisted pair	0.01	N/A
coaxial	0.012	N/A
4-conductor telephone cable	0.01	N/A
other types	0.012	N/A

SECURITY SYSTEMS *(cont.)*		
Item	**Install**	**Connect**
#14 single conductor in conduit	0.005	N/A
#16 single conductor in conduit	0.005	N/A
#18 single conductor in conduit	0.005	N/A
Miscellaneous, cable related		
shielded twisted pair termination	0.12	N/A
unshielded twisted pair termination	0.10	N/A
modular connectors, telephone	0.14	N/A
in-line couplers	0.07	N/A
male/female adapters	0.40	N/A
printer	1	.4 – 1.0
Central station equipment		
8 line receiver	3.00	4.0 – 8.0
2 line console with printout	2	1.4 – 4.5
phone line cards	0.7	.7 – 1.7
control cards	0.7	.5 – 1.5
phone line cord	0.10	N/A
50 VA power supply	0.90	.5 – 1.2
central station receiver	4.00	3.0 – 7.0
interface module	0.40	.4 – 1.1
remote control and display	0.80	.4 – 1.0

PHOTOVOLTAIC		
Item	**Ground**	**Rooftop**
Solar module, including connections		
30 cells	2.50	3.25
33 cells	2.60	3.50
36 cells	2.70	3.45
Mounting racks (module labor not included)		
1 module	3.00	3.90
2 modules	4.50	5.85
3 modules	6.00	7.80
4 modules	7.50	9.75
Tracking mounts, including controls		
1 module	4.00	5.00
2 modules	5.50	7.00
3 modules	7.00	9.00
4 modules	8.50	10.00
Concentrators (reflectors)		
for 1 module	2.40	3.20
for 2 modules	3.60	4.80
for 3 modules	4.50	6.00
for 4 modules	5.25	7.00
Power inverters		
100 watt	1.50	N/A
250 watt	1.80	N/A

PHOTOVOLTAIC *(cont.)*		
Item	**Ground**	**Rooftop**
400 watt	2.40	N/A
750 watt	3.00	N/A
1000 watt	4.00	N/A
Power disconnect switches		
2 pole, 30 amp	0.60	0.80
2 pole, 60 amp	0.75	1.00
2 pole, 100 amp	0.90	N/A
2 pole, 200 amp	1.50	N/A
3 pole, 30 amp	0.7	0.95
3 pole, 60 amp	0.9	1.20
3 pole, 100 amp	1.20	N/A
3 pole, 200 amp	1.80	N/A
Fuses, cartridge type		
0 – 30 amp	0.05	0.07
31 – 60 amp	0.05	0.07
61 – 100 amp	0.06	N/A
101 – 200 amp	0.07	N/A
Regulator		
for 1 – 4 panels	1.50	N/A
for 5 – 8 panels	2.00	N/A
for 9 – 12 panels	3.00	N/A

PHOTOVOLTAIC *(cont.)*		
Item	**Ground**	**Rooftop**
Blocking diode, for one module	0.60	0.85
Transfer switches		
2 pole, 30 amp	3.00	N/A
2 pole, 60 amp	3.25	N/A
2 pole, 100 amp	3.75	N/A
3 pole, 30 amp	3.00	N/A
3 pole, 60 amp	3.50	N/A
3 pole, 100 amp	4.00	N/A
UF cable, per foot		
#12/2 wg (with ground)	0.007	0.009
#12/3 wg	0.008	0.01
#10/1	0.008	0.01
#10/2	0.01	0.013
#10/3	0.012	0.016
#8/1	0.01	0.013
#8/2	0.011	0.014
#8/3	0.013	0.017
#6/1 AL	0.012	0.016
#6/2 AL	0.014	0.018
#6/3 AL	0.016	0.02
#4/1 AL	0.014	0.018
#2/1 AL	0.017	0.021

PHOTOVOLTAIC *(cont.)*		
Item	**Ground**	**Rooftop**
NM cables, with ground		
#14/2	0.006	0.008
#14/3	0.007	0.009
#12/2	0.007	0.009
#12/3	0.008	0.01
#10/2	0.008	0.01
#10/3	0.01	0.013
Wire terminations		
#14	0.05	0.07
#12	0.05	0.07
#10	0.06	0.08
#8	0.07	0.09
#6	0.08	0.10
#4	0.10	0.13
#2	0.12	0.16
#1	0.15	0.20
#1/0	0.18	0.24
#2/0	0.20	0.26
#3/0	0.24	0.31
#4/0	0.27	0.35

ENERGY STORAGE	
Item	
Lead-acid batteries, deep cycle, 12 volt	0.80
Battery rack materials	
galvanized unistrut, per foot	0.02
fiberglass strut, per foot	0.02
galvanized bolts	0.01
galvanized nuts	0.01
strut painting, per lineal foot	0.02
platform trays, 18" x 48"	0.50
Power inverters	
100 watt	1.50
250 watt	1.80
400 watt	2.40
750 watt	3.00
1000 watt	4.00
Power disconnect switches	
2 pole, 30 amp	0.60
2 pole, 60 amp	0.75
2 pole, 100 amp	0.90
2 pole, 200 amp	1.50
3 pole, 30 amp	0.7
3 pole, 60 amp	0.9

ENERGY STORAGE *(cont.)*	
Item	
3 pole, 100 amp	1.20
3 pole, 200 amp	1.80
Fuses, cartridge type	
0 – 30 amp	0.05
31 – 60 amp	0.05
61 – 100 amp	0.06
101 – 200 amp	0.07
Transfer switches	
2 pole, 30 amp	3.00
2 pole, 60 amp	3.25
2 pole, 100 amp	3.75
3 pole, 30 amp	3.00
3 pole, 60 amp	3.50
3 pole, 100 amp	4.00
UF cable, per foot	
#12/2 wg (with ground)	0.007
#12/3 wg	0.008
#10/1	0.008
#10/2	0.01
#10/3	0.012
#8/1	0.01

ENERGY STORAGE *(cont.)*	
Item	
#8/2	0.011
#8/3	0.013
#6/1 AL	0.012
#6/2 AL	0.014
#6/3 AL	0.016
#4/1 AL	0.014
#2/1 AL	0.017
Wire terminations	
#14	0.05
#12	0.05
#10	0.06
#8	0.07
#6	0.08
#4	0.10
#2	0.12
#1	0.15
#1/0	0.18
#2/0	0.20
#3/0	0.24
#4/0	0.27

ENERGY STORAGE *(cont.)*	
Item	
#250	0.30
#300	0.32
#350	0.34
#400	0.38
#500	0.42
Single conductors, run open, copper	
#6 THHN	0.008
#4 THHN	0.01
#3 THHN	0.012
#2 THHN	0.014
#1 THHN	0.016
#1/0 THHN	0.019
#2/0 THHN	0.022
#3/0 THHN	0.026
#4/0 THHN	0.03
#250 THHN	0.035
#300 THHN	0.039
#350 THHN	0.043
#400 THHN	0.047
#500 THHN	0.052

ENERGY STORAGE *(cont.)*	
Item	
Single conductors, run open, aluminum	
#6 THHN	0.006
#4 THHN	0.008
#3 THHN	0.01
#2 THHN	0.012
#1 THHN	0.014
#1/0 THHN	0.016
#2/0 THHN	0.018
#3/0 THHN	0.02
#4/0 THHN	0.022
#250 THHN	0.025
#300 THHN	0.028
#350 THHN	0.032
#400 THHN	0.036
#500 THHN	0.04
Single-barrel lugs	
#6	0.20
#2	0.30
#1/0	0.40
#250	0.45
#350	0.50

ENERGY STORAGE *(cont.)*	
Item	
#500	0.60
#600	0.68
#800	0.75
Double-barrel lugs	
#1/0	0.70
#250	1.00
#350	1.20
#600	1.40
#800	1.50
Three-barrel lugs	
#2/0	1.00
#250	1.50
#350	1.60
#600	2.00
#800	2.20
Four-barrel lugs	
#250	1.60
#350	2.00
#600	2.50
#800	2.70

ENERGY STORAGE *(cont.)*	
Item	
Split-bolt connectors	
#6	0.15
#4	0.30
#2	0.40
#1/0	0.50
#3/0	0.60
#250	0.80
#350	0.85
#500	0.90
#750	1.00
Miscellaneous	
tie-wraps	0.01
plywood backboard	1.50
wire markers	0.01
unistrut clamps	0.02

BUILDING AUTOMATION		
Item	**Install**	**Connect**
Control panels		
small commercial	2.00	1.0 – 2.0
medium commercial	2.50	1.5 – 3.0
large commercial/institutional	3.00	2.0 – 4.0
separate annunciator panel	1.75	1.0 – 2.0
remote controller unit	1.75	2.0 – 4.0
change battery backup	1.00	.4 – .8
replace circuit board	0.70	.4 – .8
add expansion card	0.70	1.0 – 1.9
zone expansion module	1.00	.8 – 1.6
logic module	0.90	.9 – 1.9
AC failure module	0.80	.4 – 1.0
remote status display	1	.5 – 1.5
addressable relay	0.9	1.0 – 2.0
motor protection relay	1	.1 – .3
remote status monitor	1.10	.4 – .75
trip unit	0.90	.2 – .5
timer monitor module	0.80	.4 – .8
status beeper	0.20	.1 – .2
electronic timer	0.50	.2 – .5
telephone control module	0.90	1.0 – 3.0
lightning protection module	0.80	N/A

BUILDING AUTOMATION *(cont.)*		
Item	**Install**	**Connect**
HVAC control panel	2.50	1.0 – 3.0
Remote damper	1.00	N/A
Time controls		
1 pole switch timer	0.50	N/A
2 circuit astronomical dial timer	1.20	.2 – .5
1 circuit 24 hour timer	1.00	.2 – .4
2 channel timer	1.20	.2 – .5
4 channel timer	1.35	.2 – .7
remote communicator card	0.70	.2 – .4
expansion unit	1.00	.5 – 1.5
memory storage module	0.60	.2 – .4
software	N/A	1.0 – 5.0
load scheduling timer	1.50	.5 – 1.5
special contactor	0.60	N/A
momentary contact module	0.60	N/A
photo control	0.40	N/A
Access control devices		
card reader, wall mount	0.70	.2 – 1.0
card reader, post mount	0.90	.2 – 1.0
reader controller	1.00	.4 – 1.4
palmprint identifier	2.00	2.0 – 5.0

BUILDING AUTOMATION *(cont.)*		
Item	**Install**	**Connect**
program system computer	N/A	4.0 – 10.0
telephone door security unit	4.00	5.0 – 10.0
remote intercom unit	2.00	.5 – 1.4
Powerline carrier devices		
X-10 switch modules, new work	0.25	.1 – .2
X-10 switch modules, refit	0.40	.1 – .2
X-10 receptacle module, new work	0.27	.1 – .2
X-10 receptacle module, refit	0.42	.1 – .2
X-10 chime	1.00	.1 – .2
X-10 universal module	0.40	.2 – .4
X-10 line coupler	0.50	N/A
X-10 filter	0.50	N/A
X-10 controller interface	0.70	N/A
coupling transformer	0.80	N/A
X-10 signal repeater	0.65	N/A
X-10 signal amplifier	0.65	N/A
X-10 telephone interface	0.60	.5 – 1.0
X-10 thermostat interface	0.60	.5 – 1.0
X-10 burglar alarm interface	0.70	.2 – 1.0

BUILDING AUTOMATION *(cont.)*

Item	Install	Connect
Sensors		
smoke detector	0.50	N/A
freeze sensor	0.50	N/A
water sensor	0.60	N/A
keypad	0.40	N/A
infrared motion sensor	0.50	N/A
microwave/PIR detector	0.50	N/A
driveway sensor	3.00	1.0 – 2 .0
floormat pressure sensor	0.40	N/A
sound sensor	0.40	N/A
handheld transmitter	0.20	.15 – .3
wireless money trap	0.20	.15 – .3
wall-mount transmitter	0.40	.1 – .2
cabinet tamper switch	0.30	N/A
remote switch	0.40	N/A
push button, wall mount	0.50	N/A
photocell	0.40	N/A
temperature sensor	0.40	N/A
Lighting controls		
light level sensors	0.30	N/A
lighting controller	2.50	1.0 – 3.0

BUILDING AUTOMATION *(cont.)*		
Item	**Install**	**Connect**
4 pole lighting contactor	2.00	N/A
8 pole lighting contactor	3.00	N/A
12 pole lighting contactor	4.00	N/A
2400 watt dimming controller	1.50	.2 – .5
Specific use items		
intercom, central unit	4.00	4.0 – 8.5
intercom, satellite unit	0.70	N/A
video doorphone, monitor unit	1.20	.8 – 2.0
video doorphone, outside unit	0.80	N/A
telephone-based door control	2.40	1.0 – 2.8
ceiling fan controller	1.40	N/A
floodlight/sensor unit	1.20	N/A
Miscellaneous		
console	6.00	2.0 – 8.0
key switch	1.10	N/A
door opener	2.00	N/A
door closer	2.00	N/A
electric door strike, new work	1.20	N/A
electric door strike, refit	2.20	N/A
electric mortice lock, new	2.00	N/A
electric mortice lock, refit	3.00	N/A

BUILDING AUTOMATION *(cont.)*		
Item	**Install**	**Connect**
video camera	0.5	.2 – .8
swivel camera mount	0.70	N/A
monitor	0.60	N/A
monitor mounting bracket	0.8	N/A
electronic eye, 2-piece unit	1. 2	N/A
electric gate, controls only	2.50	1.0 – 3.0
19" rack, 2'	1.80	N/A
19" rack, 4'	3.00	N/A
19" rack, 6'	4.50	N/A
tie-wraps	0.01	N/A
wire markers	0.01	N/A
plywood backboard	1.50	N/A
Cables, accessible locations		
unshielded twisted pair	0.01	N/A
shielded twisted pair	0.01	N/A
coaxial	0.012	N/A
4-conductor telephone cable	0.01	N/A
other types	0.012	N/A

BUILDING AUTOMATION *(cont.)*		
Item	**Install**	**Connect**
Miscellaneous, cable related		
shielded twisted pair terminations	0.12	N/A
unshielded twisted pair terminations	0.10	N/A
punchdown block, no terminations	1.00	N/A
wallplate with jack	0.80	N/A
modular connectors, telephone	0.14	N/A
in-line couplers	0.07	N/A
male/female adapters	0.40	N/A
coaxial crimp connectors	0.08	N/A
coaxial twist-on connectors	0.08	N/A
coaxial tee connectors	0.10	N/A
twinaxial connectors	.0.15	N/A
Computer (PC), with monitor	1.40	1.2 – 3.2
install software	N/A	1.0 – 4.0+
modular cables	0.15	N/A
cable adapters	0.35	N/A
Printers		
laser printer	1.10	.5 – 2.0
A/B printer switch, one cable	1.10	.6 – 1.5

PROGRAMMABLE CONTROLLERS		
Item	**Install**	**Connect**
Basic PLC, wall mount	2.00	4.0 – 12.0
Peripheral items		
input module, 4 point	1.00	1.0 – 4.0
input module, 8 point	1.40	2.0 – 6.0
output module, 4 point	1.00	1.0 – 4.0
output module, 8 point	2.00	2.0 – 6.0
combo module, 2 in 2 out	1.00	1.0 – 4.0
combo module, 4 in 4 out	1.40	2.0 – 6.0
combo module, 8 in 8 out	2.00	4.0 – 12.0
4 slot rack	2.00	N/A
7 slot rack	2.40	N/A
10 slot rack	2.80	N/A
13 slot rack	3.20	N/A
memory module	0.30	.8 – 1.4
power supply, 32 point	1.40	.4 – .8
power supply, 64 point	1.90	.7 – 1.4
hand-held terminal	0.50	.5 – 1.0
special purpose modules	1.00	.8 – 1.5
circuit board mounting racks	2.00	1.0 – 2.0
relay I/O board	2.00	1.4 – 2.8
32 channel I/O board	2.00	4.0 – 8.0

PROGRAMMABLE CONTROLLERS *(cont.)*		
Item	**Install**	**Connect**
48 channel I/O board	2.00	5.0 – 9.0
216 channel I/O board	3.00	10.0 – 15.0
counter/timer board	2.80	4.0 – 9.0
digital interface board	2.80	2.0 – 4.0
analog input board	2.10	.9 – 1.5
multifunction I/O card	2.70	4.0 – 9.0
multiplexer card	2.70	2.4 – 8.0
thermocouple input board	2.20	1.4 – 5.0
A/D card	2.00	1.0 – 2.4
multisensor interface card	3.00	4.0 – 12.0
universal termination board	1.40	N/A
screw terminal connection board	1.80	N/A
ribbon cable	0.25	N/A
signal conditioning backboard	3.00	N/A
signal conditioning module	0.70	1.0 – 3.0
modular motherboard	4.00	4.0 – 10.0
Computer (PC), processor only	1.00	1.0 – 3.0
Computer (PC), with monitor	1.40	1.2 – 3.2
Computer, minicomputer (AS400)	4.00	2.0 – 4.0

PROGRAMMABLE CONTROLLERS *(cont.)*		
Item	**Install**	**Connect**
Network devices		
network interface card	1.00	.5 – 1.0
host adapter, with cable	1.00	1.0 – 2.4
PAL kit	2.00	2.0 – 4.0
server disk kit	3.00	2.0 – 4.0
lap link	1.00	.7 – 1.4
LAN hub concentrator	2.00	1.9 – 4.5
multiplexer	2.40	1.9 – 4.5
telecluster	3.00	2.0 – 5.0
ethernet adapter	0.50	N/A
wireless transceiver	1.40	1.4 – 2.9
long distance adapter	1.00	.5 – 1.0
signal booster	1.00	.5 – 1.0
line splitter	1.40	1.0 – 4.0
bridge	2.00	2.0 – 7.0
router	2.00	2.0 – 7.0
WAN interface card	1.20	2.0 – 5.0
repeater	1.70	2.0 – 5.0
diagnostic modem	1.30	.9 – 3.0

PROGRAMMABLE CONTROLLERS *(cont.)*		
Item	**Install**	**Connect**
Printers		
laser printer	1.10	.5 – 2.0
Cables, per foot, accessible locations		
unshielded twisted pair	0.01	N/A
shielded twisted pair	0.01	N/A
coaxial	0.01	N/A
4-conductor telephone cable	0.01	N/A
other types	0.01	N/A
Miscellaneous, cable related		
coaxial crimp connectors	0.08	N/A
coaxial twist-on connectors	0.08	N/A
coaxial tee connectors	0.10	N/A
twinaxial connectors	0.15	N/A
shielded twisted pair terminations	0.12	N/A
unshielded twisted pair terminations	0.10	N/A
punchdown block, no terminations	1.00	N/A
wallplate with jack	0.80	N/A
modular connectors, telephone	0.14	N/A
in-line couplers	0.07	N/A

PROGRAMMABLE CONTROLLERS *(cont.)*		
Item	**Install**	**Connect**
male/female adapters	0.40	N/A
cable end assemblies	0.65	N/A
extension	0.20	N/A
Miscellaneous		
19" rack, 2'	1.80	N/A
19" rack, 4'	3.00	N/A
19" rack, 6'	4.50	N/A
tie-wraps	0.01	N/A
wire markers	0.01	N/A
plywood backboard	1.50	N/A

ROBOTS		
Item	**Install**	**Connect**
Controllers		
standard	2.00	2.5 – 5.0
high power	2.40	3.0 – 5.5
memory expansion	2.00	.8 – 3.0
teaching pendant	0.50	N/A
program backup interface	0.50	N/A
RS 232 adapter	0.40	N/A
controller cable	0.40	N/A
5 meter controller cable	0.50	N/A
inter-axis cable	0.40	N/A
connector covers	0.40	N/A
mounting brackets	0.40	N/A
guide rail, 500mm	1.00	N/A
guide rail, 600mm	1.25	N/A
guide rail, 800mm	1.50	N/A
guide rail, 1000mm	2.00	N/A
relay I/O board	2.00	1.4 – 2.8
32 channel I/O board	2.00	4.0 – 8.0
48 channel I/O board	2.00	5.0 – 9.0
216 channel I/O board	3.00	10.0 – 15.0
counter/timer board	2.80	4.0 – 9.0

ROBOTS *(cont.)*		
Item	**Install**	**Connect**
digital interface board	2.80	2.0 – 4.0
analog input board	2.10	.9 – 1.5
multifunction I/O card	2.70	4.0 – 9.0
multiplexer card	2.70	2.4 – 8.0
thermocouple input board	2.20	1.4 – 5.0
A/D card	2.00	1.0 – 2.4
multisensor interface card	3.00	4.0 – 12.0
universal termination board	1.40	N/A
screw terminal connection board	1.80	N/A
ribbon cable	0.25	N/A
signal conditioning backboard	3.00	N/A
signal conditioning module	0.70	1.0 – 3.0
modular motherboard	4.00	4.0 – 10.0
Actuators		
8Kg load, 300mm	4.00	4.0 – 8.0
8Kg load, 400mm	4.50	4.0 – 8.0
8Kg load, 500mm	5.00	4.0 – 8.0
12Kg load, 300mm	5.00	4.0 – 8.0
12Kg load, 400mm	5.50	4.0 – 8.0
12Kg load, 500mm	6.00	4.0 – 8.0
15Kg load, 300mm	6.00	4.0 – 8.0

ROBOTS *(cont.)*		
Item	**Install**	**Connect**
15Kg load, 500mm	6.50	4.0 – 8.0
vertical, 10Kg, 200mm	5.00	4.0 – 8.0
vertical, 10Kg, 400mm	5.75	4.0 – 8.0
vertical, 10Kg, 600mm	6.50	4.0 – 8.0
XY combo, light duty, 300mm	7.00	6.0 – 10.0
XY combo, light duty, 400mm	8.00	6.0 – 10.0
XY combo, light duty, 500mm	9.00	6.0 – 10.0
XY combo, light duty, 600mm	10.00	6.0 – 10.0
XY combo, heavy duty, 300mm	7.50	6.0 – 10.0
XY combo, heavy duty, 400mm	8.50	6.0 – 10.0
XY combo, heavy duty, 500mm	9.50	6.0 – 10.0
XY combo, heavy duty, 600mm	10.50	6.0 – 10.0
XZ combo, 2Kg, 300mm	8.00	6.0 – 10.0
XZ combo, 2Kg, 400mm	9.00	6.0 – 10.0
XZ combo, 2Kg, 500mm	10.00	6.0 – 10.0
XZ combo, 2Kg, 600mm	11.00	6.0 – 10.0
XZ combo, 4Kg, 300mm	9.00	6.0 – 10.0
XZ combo, 4Kg, 400mm	10.00	6.0 – 10.0
XZ combo, 4Kg, 500mm	11.00	6.0 – 10.0
XZ combo, 4Kg, 600mm	12.00	6.0 – 10.0
Computer (PC), processor only	1.00	1.0 – 3.0

ROBOTS *(cont.)*		
Item	**Install**	**Connect**
Computer (PC), with monitor	1.40	1.2 – 3.2
Computer, minicomputer (AS400)	4.00	2.0 – 4.0
Network devices		
network interface card	1.00	.5 – 1.0
host adapter, with cable	1.00	1.0 – 2.4
PAL kit	2.00	2.0 – 4.0
server disk kit	3.00	2.0 – 4.0
lap link	1.00	.7 – 1.4
LAN hub concentrator	2.00	1.9 – 4.5
multiplexer	2.40	1.9 – 4.5
telecluster	3.00	2.0 – 5.0
ethernet adapter	0.50	N/A
wireless transceiver	1.40	1.4 – 2.9
long distance adapter	1.00	.5 – 1.0
signal booster	1.00	.5 – 1.0
line splitter	1.40	1.0 – 4.0
bridge	2.00	2.0 – 7.0
router	2.00	2.0 – 7.0
WAN interface card	1.20	2.0 – 5.0
repeater	1.70	2.0 – 5.0
diagnostic modem	1.30	.9 – 3.0

ROBOTS *(cont.)*

Item	Install	Connect
Cables, per foot, accessible locations		
unshielded twisted pair	0.01	N/A
shielded twisted pair	0.01	N/A
coaxial	0.01	N/A
4-conductor telephone cable	0.01	N/A
other types	0.01	N/A
Miscellaneous, cable related		
coaxial crimp connectors	0.08	N/A
coaxial twist-on connectors	0.08	N/A
coaxial tee connectors	0.10	N/A
twinaxial connectors	0.15	N/A
shielded twisted pair terminations	0.12	N/A
unshielded twisted pair terminations	0.10	N/A
punchdown block, no terminations	1.00	N/A
wallplate with jack	0.80	N/A
modular connectors, telephone	0.14	N/A
in-line couplers	0.07	N/A
male/female adapters	0.40	N/A
cable end assemblies	0.65	N/A
extension	0.20	N/A

ROBOTS *(cont.)*		
Item	**Install**	**Connect**
Miscellaneous		
19" rack, 2'	1.80	N/A
19" rack, 4'	3.00	N/A
19" rack, 6'	4.50	N/A
tie-wraps	0.01	N/A
wire markers	0.01	N/A
plywood backboard	1.50	N/A

CHAPTER 4
Technology Unit Pricing Forms

Unit pricing for technology installations is essentially the same as for the residential and commercial unit pricing described in **Chapter 2**. All of the same cautions must be observed, and all of the same benefits stand to be gained. Unit pricing done right is fast, easy and inexpensive. Just make sure you apply the technique thoughtfully.

This chapter contains the most commonly used unit prices from a variety of specialties. We have purposefully left out material prices and labor hours. You should fill these out yourself, and think about each unit carefully as you do so. Exactly what work is involved? How will these devices interface with other items? Will you have any trouble obtaining some of the components? Give each unit your full attention as you prepare the form. Do it right the first time.

Note that the quantities for cable, conduit and wire are averages. These figures were obtained from the estimates and job accounting records of actual installations. Certainly your jobs may vary, and you may wish to modify these numbers a bit. The numbers you find here should, however, be reasonably close to your actual installed quantities.

SECURITY SYSTEM, HOUSE

Quantity	Item	Mat.	Ext.	Hrs.	Ext.
1	Control panel, includes backup power				
1	Interior siren				
3	Door contacts				
8	Surface mount glass break detectors				
6	Window screen sensors				
1	Outdoor siren				
2	Passive infrared sensors				
1	Smoke/heat detector				
1	Keypad				
700'	#22 cable				
	Material cost				

	Sales tax, misc.				
	Material total				
	hrs. @				
	Total cost				
	50% O & P				
	Selling price — $				

SECURITY SYSTEM, LARGE HOUSE

Quantity	Item	Mat.	Ext.	Hrs.	Ext.
1	Control panel, includes backup power				
1	Interior siren				
5	Door contacts				
12	Surface mount glass break detectors				
15	Window screen sensors				
1	Outdoor siren				
5	Passive infrared sensors				
2	Smoke/heat detector				
2	Keypad				
1500'	#22 cable				
	Material cost				

	Sales tax, misc.				
	Material total				
	hrs. @				
	Total cost				
	50% O & P				
	Selling price — $				

SECURITY SYSTEM, APARTMENT

Quantity	Item	Mat.	Ext.	Hrs.	Ext.
1	Control panel, includes backup power				
1	Interior siren				
1	Smoke/heat detector				
2	Surface mount glass break detectors				
2	Door contacts				
1	Keypad				
250'	#22 gauge cable				
	Material cost				
	Sales tax, misc.				
	Material total				
	hrs. @				

	Total cost				
	50% O & P				
	Selling price — $				

WIRELESS SECURITY SYSTEM, HOUSE

Quantity	Item	Mat.	Ext.	Hrs.	Ext.
1	Control panel, includes backup power				
1	Interior siren				
10	Window/door sensor/transmitters				
2	Passive infrared sensor/transmitters				
5	Receivers				
1	Outdoor siren				
Lot	Wiring for outdoor siren				
1	Wireless smoke detector				
1	Remote keypad				
2	Panic transmitters				
	Material cost				

	Sales tax, misc.				
	Material total				
	hrs. @				
	Total cost				
	50% O & P				
	Selling price — $				

WIRELESS SECURITY SYSTEM, APARTMENT

Quantity	Item	Mat.	Ext.	Hrs.	Ext.
1	Control panel, includes backup power				
1	Interior siren				
3	Window/door sensor/transmitters				
1	Passive infrared sensor/transmitters				
2	Receivers				
1	Panic transmitter				
1	Wireless smoke detector				
	Material cost				
	Sales tax, misc.				
	Material total				
	hrs. @				

	Total cost				
	50% O & P				
	Selling price — $				

PASSIVE INFRARED SENSOR

Quantity	Item	Mat.	Ext.	Hrs.	Ext.
1	Passive infrared motion sensor				
50'	#22/4 cable				
	Material cost				
	Sales tax, misc.				
	Material total				
	hrs. @				
	Total cost				
	50% O & P				
	Selling price — $				

SURFACE MOUNT GLASS-BREAK DETECTOR

Quantity	Item	Mat.	Ext.	Hrs.	Ext.
1	Surface mount glass break detector				
30'	#22 cable				
	Material cost				
	Sales tax, misc.				
	Material total				
	hrs. @				
	Total cost				
	50% O & P				
	Selling price — $				

ACCESS CONTROL, SMALL OFFICE

Quantity	Item	Mat.	Ext.	Hrs.	Ext.
1	Keypad, weatherproof				
2	Magnetic card readers				
3	12-volt locks, includes strike plate,				
	accessories				
1	Reader controller				
20	Magnetic cards				
1	Programming tool				
200'	1/2" EMT				
14	1/2" EMT connector				
20	1/2" EMT couplings				
22	1/2" straps				
250'	#22 cable				

	Material cost				
	Sales tax, misc.				
	Material total				
	hrs. @				
	Total cost				
	50% O & P				
	Selling price — $				

WINDOW SCREEN SENSOR

Quantity	Item	Mat.	Ext.	Hrs.	Ext.
1	Window screen sensor				
30'	#22 cable				
	Material cost				
	Sales tax, misc.				
	Material total				
	hrs. @				
	Total cost				
	50% O & P				
	Selling price — $				

ACCESS CONTROL, OFFICE BUILDING

Quantity	Item	Mat.	Ext.	Hrs.	Ext.
4	Keypads, weatherproof				
12	Magnetic card readers				
16	12-volt locks, includes strike plate,				
	accessories				
4	Reader controllers				
1	Software (for use with existing PC)				
Lot	Computer connection hardware				
1	Printer				
100	Magnetic cards				
1750'	1/2" EMT				
72	1/2" EMT connectors				
180	1/2" EMT couplings				

200	1/2" straps				
2000'	#22 cable				
	Material cost				
	Sales tax, misc.				
	Material total				
	hrs. @				
	Total cost				
	50% O & P				
	Selling price — $				

CCTV MONITORING #1

Quantity	Item	Mat.	Ext.	Hrs.	Ext.
14	Black & white camera, auto-iris,				
	with lens				
14	Mounts				
2	Camera housings				
4	Outdoor housings				
2	Dummy cameras				
1	Time lapse video recorder				
2	8-position sequencer				
4	12" black & white monitors				
800'	RG-59U coaxial cable				
	Material cost				

	Sales tax, misc.				
	Material total				
	hrs. @				
	Total cost				
	50% O & P				
	Selling price — $				

CCTV MONITORING #2

Quantity	Item	Mat.	Ext.	Hrs.	Ext.
8	Black & white camera, auto-iris,				
	with lens				
8	Mounts				
1	Camera housings				
1	12" black & white monitor				
1	8-position sequencer				
1	Time lapse video recorder				
500'	RG-59U coaxial cable				
	Material cost				
	Sales tax, misc.				
	Material total				

	hrs. @				
	Total cost				
	50% O & P				
	Selling price — $				

SECURITY CAMERA

Quantity	Item	Mat.	Ext.	Hrs.	Ext.
1	Black & white camera, auto-iris,				
	with lens				
1	Mount				
110'	RG-59U coaxial cable				
2	Weatherproof boxes				
1	Weatherproof duplex				
	receptacle cover				
1	Blank cover				
1	Duplex receptacle				
110'	1/2" EMT				
2	1/2" EMT connectors				
11	1/2" EMT couplings				

12	½" straps				
240'	#12 THHN				
1	Cable connector				
	Material cost				
	Sales tax, misc.				
	Material total				
	hrs. @				
	Total cost				
	50% O & P				
	Selling price — $				

HOME SOUND SYSTEM

Quantity	Item	Mat.	Ext.	Hrs.	Ext.
1	Built-in radio/cassette unit				
1	Power feed				
8	8" High fidelity speakers,				
	rough-in brackets				
4	Volume controls, w/impedance				
	matching				
250'	16 gauge speaker wire				
	Material cost				
	Sales tax, misc.				
	Material total				
	hrs. @				

	Total cost				
	50% O & P				
	Selling price — $				

BUILT-IN SPEAKER

Quantity	Item	Mat.	Ext.	Hrs.	Ext.
1	8" High fidelity speaker,				
	rough-in bracket				
32'	16 gauge speaker wire				
	Material cost				
	Sales tax, misc.				
	Material total				
	hrs. @				
	Total cost				
	50% O & P				

Selling price — $

HOME THEATRE SYSTEM

Quantity	Item	Mat.	Ext.	Hrs.	Ext.
1	Large-screen television				
1	THX amp and controller				
4	Wall speakers				
2	Sub-woofers				
1	VCR				
180'	16 gauge speaker wire				
1	Dedicated power circuit				
	Material cost				
	Sales tax, misc.				
	Material total				
	hrs. @				

	Total cost				
	30% O & P				
	Selling price — $				

HOME INTERCOM #1

Quantity	Item	Mat.	Ext.	Hrs.	Ext.
1	Master station w/radio,				
	cassette, auxiliary, clock				
5	Indoor speaker stations				
1	Outdoor speaker station				
2	Door speaker stations				
500'	6-conductor cable				
1	Power feed				
	Material cost				
	Sales tax, misc.				
	Material total				
	hrs. @				

	Total cost				
	50% O & P				
	Selling price — $				

HOME INTERCOM #2

Quantity	Item	Mat.	Ext.	Hrs.	Ext.
1	Master station w/radio,				
	cassette, auxiliary, clock				
5	Indoor speaker stations				
1	Outdoor speaker station				
2	Door speaker stations				
2	8" Indoor speakers				
2	8" Outdoor speakers				
1	Indoor remote control				
1	Outdoor remote control				
800'	6-conductor cable				
1	Power feed				

	Material cost				
	Sales tax, misc.				
	Material total				
	hrs. @				
	Total cost				
	50% O & P				
	Selling price — $				

VIDEO DOOR SYSTEM

Quantity	Item	Mat.	Ext.	Hrs.	Ext.
1	Master station w/radio and monitor				
5	Indoor speaker stations				
1	Outdoor speaker station				
2	Door camera and speaker stations				
2	Indoor camera stations				
200'	Coaxial cable, RG-59U				
500'	6-conductor cable				
1	Power feed				
2	Electric door locks, strike plate, etc.				
2	12 volt power feeds				
	Material cost				

	Sales tax, misc.				
	Material total				
	hrs. @				
	Total cost				
	50% O & P				
	Selling price — $				

INTERCOM STATION

Quantity	Item	Mat.	Ext.	Hrs.	Ext.
1	Indoor speaker station				
50'	6-conductor cable				
	Material cost				
	Sales tax, misc.				
	Material total				
	hrs. @				
	Total cost				
	50% O & P				
	Selling price — $				

FIRE ALARM #1

Quantity	Item	Mat.	Ext.	Hrs.	Ext.
1	Fire alarm panel				
4	Pull stations				
12	Smoke alarms				
1	Duct detector				
2	End-of-line (EOL) resistors				
1	Dedicated circuit power feed				
50'	#22/4 phone cable				
1400'	#18 TFN wire				
500'	½" EMT				
34	½" EMT connectors				
50	½" EMT couplings				
55	½" EMT straps				

7	4" square boxes				
3	4" blank covers				
12	ceiling boxes with hanger				
	Material cost				
	Sales tax, misc.				
	Material total				
	hrs. @				
	Total cost				
	50% O & P				
	Selling price — $				

FIRE ALARM #2

Quantity	Item	Mat.	Ext.	Hrs.	Ext.
1	Fire alarm panel				
15	Pull stations				
30	Smoke alarms				
2	Duct detector				
8	End-of-line (EOL) resistors				
1	Dedicated circuit power feed				
50'	#22/4 phone cable				
3900'	#18 TFN wire				
1400'	½" EMT				
94	½" EMT connectors				
140	½" EMT couplings				
95	½" EMT straps				

20	4" square boxes				
8	4" blank covers				
30	Ceiling boxes with hanger				
	Material cost				
	Sales tax, misc.				
	Material total				
	hrs. @				
	Total cost				
	35% O & P				
	Selling price — $				

PULL STATION

Quantity	Item	Mat.	Ext.	Hrs.	Ext.
1	Pull station				
120'	#18 TFN wire				
30'	½" EMT				
2	½" EMT connectors				
3	½" EMT couplings				
3	½" EMT straps				
	Material cost				
	Sales tax, misc.				
	Material total				
	hrs. @				

	Total cost				
	50% O & P				
	Selling price — $				

SMOKE DETECTOR

Quantity	Item	Mat.	Ext.	Hrs.	Ext.
1	Smoke detector				
65'	#18 TFN wire				
30'	½" EMT				
2	½" EMT connectors				
3	½" EMT couplings				
3	½" EMT straps				
1	Ceiling box with hanger				
	Material cost				
	Sales tax, misc.				
	Material total				
	hrs. @				

	Total cost				
	50% O & P				
	Selling price — $				

DUCT DETECTOR

Quantity	Item	Mat.	Ext.	Hrs.	Ext.
1	Duct detector				
160'	#18 TFN wire				
40'	½" EMT				
2	½" EMT connectors				
4	½" EMT couplings				
5	½" EMT straps				
	Material cost				
	Sales tax, misc.				
	Material total				
	hrs. @				

	Total cost				
	50% O & P				
	Selling price — $				

FIBER TERMINATION #1

Quantity	Item	Mat.	Ext.	Hrs.	Ext.
1	ST connector				
	Sanding pads, pucks, adhesive, etc.				N/A
	Testing		N/A		
	Material cost				
	Sales tax, misc.				
	Material total				
	hrs. @				

	Total cost				
	40% O & P				
	Selling price — $				

FIBER TERMINATION #2

Quantity	Item	Mat.	Ext.	Hrs.	Ext.
½	Factory single mode				
	jumper (cut in field)				
2	End terminations		N/A		
	Sanding pads, pucks, adhesive, etc.				N/A
1	Mechanical splice				
	Testing		N/A		
	Material cost				

	Sales tax, misc.				
	Material total				
	hrs. @				
	Total cost				
	40% O & P				
	Selling price — $				

ZONE CONTROL SYSTEM

Quantity	Item	Mat.	Ext.	Hrs.	Ext.
4	Electric duct dampers				
200'	#22 cable				
160'	$\frac{1}{2}$" EMT				
8	$\frac{1}{2}$" EMT connectors				
16	$\frac{1}{2}$" EMT couplings				
20	$\frac{1}{2}$" straps				
2	Thermostats				

	Material cost				
	Sales tax, misc.				
	Material total				
	hrs. @				
	Total cost				
	50% O & P				
	Selling price — $				

CEILING FAN

Quantity	Item	Mat.	Ext.	Hrs.	Ext.
1	52" Ceiling fan				
150'	#12 THHN				
60'	1/2" EMT				
2	1/2" EMT connectors				
6	1/2" EMT couplings				
7	1/2" straps				
1	4" Square box, with round cover				

	Material cost				
	Sales tax, misc.				
	Material total				
	hrs. @				
	Total cost				
	50% O & P				
	Selling price — $				

TECHNOLOGY PRICING NOTES

CHAPTER 5
Algebraic Estimating

ALGEBRAIC ESTIMATING

Algebraic estimating is a relatively new and different method of electrical estimating that offers immense time savings for estimating commercial, industrial and institutional electrical projects.

The algebraic method works by the intersectional relationships that are inherent in any electrical project. It makes use of the intersectional proportionality fact by allowing one section of the project to be calculated with a mathematic formula, rather than by counting and tabulating data. Careful application and a few calculations are required, but the final result is a good estimate in a fraction of the time normally spent.

This is how algebraic estimating works:

1. **A takeoff must be done of all the determining sections of the job.**
2. **These determining sections must be priced and labored, just as in the itemizing method.**
3. **Proportionalities and prices must be calculated for these sections.**
4. **These proportions and prices must then be adjusted by comparison with a similar job.**
5. **A preliminary total cost is calculated.**
6. **This preliminary total cost (or preliminary bid price) must be adjusted to determine a final bid price (selling price).**

Algebraic estimating is set up so that one section of the job will be calculated by comparison to other sections, rather than being taken off. This is the "miscellaneous material" section — the everyday types of materials that are used on virtually every job. This includes branch circuits, wiring devices, and so on. Some typical "miscellaneous material" items are:

Conduit (all types)
Underfloor and wall duct
Wireways and gutters
Cable tray
Wiremold
Outlet boxes
Condulets
Enclosures and cabinets
Unistrut
Terminals and lugs
Fittings (all types)
Straps (all types)
Fasteners (all types)
Grounding clamps and pigtails
Ground rods
Wire and cable (all types)
Receptacles (all types)
Switches (all types)
Cord and cord sets
Plaster rings
Extensions
Finish plates (all types)

LIGHT FIXTURES

Light fixtures account for 30 to 40 percent of the cost of an average commercial electrical project, making them the single most important section in the average project. This section of the job typically includes the following types of materials:

Light fixtures (all types)

Fixture hangers

Lighting track

Stems and chains

Remote ballasts

Egg crate louvers

Exit signs

Emergency lights

Lamps (bulbs) (all types)

Included in this section are all types of interior light fixtures and their accessories. Security lights that are mounted on outside walls are also included in this section, but parking lot lighting is not. Parking lot lighting must be included in the "special system" section, as parking lot lighting is not consistent from job to job. Putting it in the light fixture section throws the system off on certain jobs.

Typically, the light fixtures are bid to contractors as a package. That is, a supplier gets the quantities of all the various fixture types from the estimator, then he or she talks to selected lighting reps (and factories, if necessary) and works up one price for the whole lot of light fixtures.

LIGHT FIXTURES *(cont.)*

All that is necessary for this section is to count all the light fixtures on the particular job, assign each fixture a labor unit and extend and total the labor hours allotted for the installation of the fixtures. The estimator notifies one or more suppliers of the fixture quantities and they provide a price.

Earlier in this chapter we said that the miscellaneous materials section was almost completely determined by the other sections. As we look at light fixtures, this statement should become a littler clearer.

Light fixtures are the single largest load on almost any electrical system, thus their number largely determines how much branch circuitry will be needed in the job. Every light fixture needs power to operate, so each fixture will need a branch circuit running to it. For a certain number of branch circuits, a subpanel and feeder will be required.

Thus we see that the amount of miscellaneous wiring materials on a typical project is greatly determined by the amount of lighting in the project. We can say that they are directly proportional to each other. When one increases, the other will increase accordingly; and when one decreases, the other will decrease accordingly.

DISTRIBUTION EQUIPMENT

The distribution equipment section of an electrical project includes all the equipment that is used to distribute and control electrical power. Typical items of distribution equipment are:

Switchboards

Panelboards

Load centers

Motor control centers

Motor starters

Circuit breakers

Fuses

Meters

Meter centers

Current transformer cabinets

Transformers

Capacitors

Enclosed circuit breakers

Safety switches

Manual starters

Lighting contactors

Bus duct

Time clocks

Push-button stations

Other control items

These are the standard items that are considered to be included in distribution equipment. However, this section does not include large items like substations and overhead distribution. These and similar items are included in the special systems section of the project.

DISTRIBUTION EQUIPMENT *(cont.)*

Distribution equipment makes up one large section of the typical project — about 12 to 20 percent. However, in actual practice the section is broken down into two separate parts.

Virtually all suppliers and manufacturers treat the one-line diagram as a part by itself. Typically, the various distribution equipment manufacturers (e.g., Square D, G.E., Westinghouse, and I.T.E.) have representatives in every major marketplace who handle all the jobs that go out for bid in their areas. These representatives make it their business to take off every job in their areas and see to it that all their distributors are furnished with a price prior to the bid time. When these representatives take off the job, they usually do the one-line diagram only.

The control items and any other power items that don't appear on the one-line diagram comprise the second part of distribution equipment. These ordinarily need to be taken off by the estimator and then priced and labored along with all the other materials. Motor starters, push-button stations, special contactors, relays, safety switches, enclosed circuit breakers, pressure switches and time clocks are some examples of items considered in this group.

For the algebraic estimating method, only the one-line diagram part of distribution equipment is used, not the other control items. This part is typically called the "switchgear," or "gear" for short. Using the switchgear (as shown on the one-line diagram) gives some major advantages over using the whole section of distribution equipment. As with lighting fixtures, you can get someone else to do all your pricing for you. The manufacturer's representative does a takeoff of the job and relays a price to your distributor, who in turn calls you up with a bid. All your pricing and extensions are already done for you.

The switchgear is the only part of the distribution equipment that needs to be used because our concern is with those sections that dictate (or determine) the extent of the other sections of the job, particularly miscellaneous materials. The switchgear does this as well as or better than the whole distribution equipment section. The switchgear carries the whole electrical load for the project, so if there were more equipment going into the project, there would be more switchgear. As with lighting fixtures, the miscellaneous material section is directly proportional to the switchgear.

The switchgear is a perfect indicator of the extent of the miscellaneous materials that will go into the job. In addition, the one-line diagram is about the easiest thing there is for an estimator to take off. All the panels, transformers, switches, and so on, are shown in one diagram; there is no looking through the other sheets for distribution equipment that is necessary.

As we have seen, there are several advantages to using the light fixtures and switchgear sections. These are two of the easiest sections of the job to take off (especially the switchgear). They are the best indicators of the amount of miscellaneous materials that will be required and, further, someone else does all your material pricing for you. That's pretty tough to beat!

SPECIAL SYSTEMS

The last section used for the algebraic estimating method is special systems. Special systems include:

Fire alarm systems

Nurse call systems

Sound systems

TV systems

Security systems

Special hospital systems

Emergency power systems

UPS systems

Substations

Overhead transmission

Lightning protection systems

Signaling systems
(e.g., bells, buzzer, chimes)

Electric heating

Special systems are various electrical systems that provide some specialized function and are typically auxiliary to the basic building wiring system.

The special systems section of the project is not used as a determining section but must be included in the algebraic method, as it completes the scope of the work. Because of the way that special systems vary from job to job, we handle special systems differently than we do any other section.

Special systems are treated as a separate entity from the rest of the job. The algebraic method has been designed so that special systems can be adjusted up or down independently on the other sections. This is necessary because special systems have no consistent pattern that they follow for any particular type of job. One job may have a fire alarm system; another job may have a fire alarm system, a sound system, and lightning protection, and a third job may have no special systems at all.

The ability to handle special systems separately is essential for the success of the algebraic method. Fortunately, this is fairly easy to do. The details will be explained later.

AREA

The next major factor in an algebraic estimate is the area of the job; that is, the physical size of the job. In the United States this means square footage. In most countries, it means square meters.

The system doesn't have to be set up this way. Proportions only could be used, still giving the same results, but the feel of the system wouldn't be the same. Tying the system to square footage makes the system a lot more understandable for almost everyone. It puts all the numbers and calculations into a form that is easier to comprehend.

As you will notice, the algebraic system often expresses area not only in square feet but "per thousand square feet." The abbreviation "/MSF" is used for "per thousand square feet." Per thousand square feet is used rather than per square foot because the numbers are a lot easier to work with. If you use per square foot only, the numbers have many decimal places and errors in calculation are much more likely. For the sake of ease of use and freedom from error, per thousand square feet is preferable.

PROPORTIONALITY

Here are the basics of proportionality, with some examples given for clarity.

The amount of miscellaneous wiring materials is completely dependent upon the amount of items that are served. Every item that uses electrical power needs power brought to it by means of miscellaneous wiring materials. It is for this reason that light fixtures and switchgear are used as determining sections.

Why? First, because light fixtures are the primary load on the system; that is, there are more light fixtures that need to get wired than any other item. Second, because all the load flows through the switchgear. If there is more load, that means more miscellaneous wiring materials, and it also means more switchgear. Less load means less wiring materials and less switchgear.

Here is an example of how this works. Let's say that we just completed an elementary school. The cost of light fixtures on the job came out to $1,200.00 per thousand square feet. The number of hours needed to install these light fixtures came out to 24 per thousand square feet. The cost of switchgear came out to $550.00 per thousand square feet, and the number of hours needed to install this switchgear came out to 7 hours per thousand square feet. We add up the cost of all the miscellaneous wiring materials and find that they came out to $1,800.00 per thousand square feet and took 150 hours per thousand square feet to install.

We have a pretty good idea of what our costs really were on this job. Now let's say that a couple of months later another elementary school comes up for bid.

PROPORTIONALITY *(cont.)*

After doing just part of the takeoff, we find the following: The cost of light fixtures comes out to $1,320.00/MSF (per thousand square feet). The number of hours needed to install these light fixtures comes out to between 26 and 27 hours/MSF. The cost of switchgear comes out to $605.00/MSF, and the number of hours needed to install this switchgear comes out to 7.7 hours/MSF.

After looking at these numbers, we find out that the cost of light fixtures is about 10 percent higher on the new job than it was on the completed job. Then we find out that the number of hours needed to install these fixtures is also about 10 percent higher, as is the installation of the switchgear. By our calculations we find out that there is 10 percent more load per square foot in the new job and also 10 percent more labor will be needed per square foot. What is going to happen to our miscellaneous wiring material? Simple, it is going to go up 10 percent per square foot too.

Of course this example is very simplistic, but it illustrates the point well: miscellaneous wiring materials are directly proportional to the amount of load. You also notice that we are comparing a school to a school. This is essential as different types of jobs have different characteristics and these figures cannot be transferred from one type of job to another.

For every job that is estimated using the algebraic estimating method, an appropriate comparison job is needed for working out all the calculations. In order to determine the proper proportions of a job, a good indicator is needed; that is, you need to know the relationships (ratios) of one section to another for this type of job.

The important thing to remember about using comparison jobs is that they must be good comparisons. You can't compare a warehouse to an office building. As they say, "You have to compare apples to apples."

CONSISTENT RELATIONSHIPS

The fact that really makes the system work is this: the relationships of the three sections we have discussed — light fixtures, switchgear and miscellaneous wiring materials — are almost exactly the same for different jobs of the same type. For example, the relationship of these three sections will be virtually identical from one high school to another high school. If the ratio is 12:7:15 on one supermarket job, you can be sure that the ratio will be almost exactly the same for a similar supermarket.

Given this fact, we are able to take advantage of these consistent proportions with a handful of formulas and a partial takeoff.

ALGEBRA

The algebraic estimating method uses one basic algebraic equation. (The term "algebraic equation" may sound hard, but don't let it scare you. The guy who figured out the formula got a D in first-year algebra.) This is the formula we use to get a balanced multiplier:

$$\frac{(A \times B) + C}{A + 1} = \text{Balanced multiplier}$$

For example, let's say A = 5, B = 20 and C = 7. Inserting these figures in the formula, we have

$$\frac{(5 \times 20) + 7}{5 + 1}$$

Performing the multiplication in the dividend and the addition in the divisor gives us

$$\frac{100 + 7}{6}$$

Complete the operations:

$$\frac{107}{6} = 17.83$$

What this formula does is to give more or less weight to the larger or smaller parts of the job. This is called a balanced multiplier.

THE TAKEOFF

The process of taking off a set of plans using the algebraic estimating method is similar in some respects to the standard itemized method, but there are some important differences.

The first (and most important) difference is that only certain parts of the job are taken off rather than the entire project down to the smallest locknut. This is really where the great savings in time come from. The one greatest advantage of this system is that it permits you to do only a partial takeoff and yet obtain the right price for the project.

The second difference is that a little extra effort is needed to review the job. This extra effort in review should be given all the time, whatever method of estimating is used. But as most estimators are pressed too hard for time, they often shorten the amount of time they spend reviewing the project. When using the algebraic estimating method, you should not shorten this step. Keep in mind that you will not be spending anywhere near the amount of time looking at the plans as you did with the itemized method.

A third difference is that you will also need to do a takeoff of the project's area. Because the system is area based, it retains the same type of feel as regular electrical estimating. If only a proportionally based system were used, a lot of people would be uncomfortable with it. While a system that used mathematics only would have certain advantages, it would be better suited to mathematicians than to electrical estimators, so we will stick to the area-based system and do a takeoff of the area (square footage).

Here are the steps in doing an algebraic estimating method takeoff:

1. Job review.

2. Takeoff of light fixtures.

THE TAKEOFF *(cont.)*

3. Takeoff of one-line diagram.

4. Takeoff of area.

5. Special systems.

6. Final review.

As you look over these six steps, you will notice that there is only one step in the takeoff that is not usually included in conventional estimating: the area takeoff. All the other steps should be quite familiar to any experienced electrical estimator.

The takeoffs of the light fixtures, one-line diagram and special systems are exactly the same as in the itemized method. The job review is also very similar to what should be done in the itemized method.

JOB REVIEW

Reviewing the job before doing a takeoff is not unique to the algebraic estimating method and is certainly nothing new. The first thing to look over is the spec book. You need to know the rules of the project. The first section of the specs, typically known as the "General Conditions," contains the rules that spell out exactly how the job will be run.

Next you should review the sections of the specs dealing with the structural and architectural parts of the job. You really don't need to spend too much time going over these sections, but you should double-check to make sure that there is nothing in these sections that affects you.

Having reviewed the specs, the next step is to review the drawings. First, look over the site plan to see

JOB REVIEW *(cont.)*

if there is any of your work shown there. (Often you will find site lighting and service runs on the site plan.)

Next, review the structural and architectural sheets to familiarize yourself with how the job will actually be built. Look over the various wall sections and details. Keep your eyes open to see if there are any stray light fixtures or other electrical items that turn up on these drawings. Look for any details of the electrical installation.

Take a look through the mechanical drawings to see what type of mechanical equipment will be used and where and how it will be mounted. There may also be motor schedules in this section that you should be familiar with.

Now you come to the electrical drawings. Look them over carefully at this stage of your takeoff. Watch for anything that is left off the drawings or that is unclear. Try to get a feel for just how big the job is and how it will be put together.

Decide whether you will need any specialty subcontractors. If you do, call them up and get them started figuring the job. Talk to your suppliers and get them started on figuring the light fixtures and switchgear. If there are any other special types of quotes that you will need to estimate the job properly, be sure to get someone started working on them.

LIGHT FIXTURE TAKEOFF

The takeoff of the light fixtures is pretty straightforward. You need to get a complete count of each individual type of light fixture. You then will need to

LIGHT FIXTURE TAKEOFF *(cont.)*

labor all the light fixtures and obtain a figure of total hours that will be required to install them.

Always watch for any special hangers, stems, louvers, and so on, and include these items in your figures. Make sure you take into account such things as ceiling height and ceiling type when you apply labor units to the light fixtures. Also make sure that you include the cost of lamps (bulbs).

You should not include the cost of labor for the site lighting along with that for the light fixtures. The site lighting (if any) will be included with special systems. If you would like to take off the site lighting while you are doing the regular light fixtures (since you will already have the plans open right in front of you and a pencil in your hand), that would be a good idea, but be sure to keep the two categories separate. (Remember also that security-type light fixtures mounted on the outside walls of the building should be included with the regular light fixtures, not with the site lighting.)

SWITCHGEAR TAKEOFF

The next step is to do a takeoff of the one-line diagram. Often one of the switchgear manufacturers will furnish you with a copy of their takeoff of the one-line diagram, but you should do your own takeoff anyway as it only takes a few minutes, and you should know how all the distribution equipment fits together.

Count all the items shown on the one-line diagram. The one-line diagram typically includes items such as main switchboards, distribution panelboards, power panels, lighting panels, transformers and meters.

SWITCHGEAR TAKEOFF *(cont.)*

Generally feeders are also shown on the one-line diagram. It is not necessary for you to take off these feeders, too, as the system takes care of the feeders "automatically." (Feeders are proportional to the switchgear; 200 amps worth of switchgear will require 200 amps worth of feeders, etc.) Feeders come under the "miscellaneous wiring materials" section of the job.

Do a takeoff of the one-line diagram and labor all the individual items to obtain a figure of total labor hours that will be required for the installation of all the switchgear shown on the one-line diagram.

Every now and then you will find that an engineer has put a lot more on the one-line diagram than anyone else does. If this happens, use only the items that normally are included on the one-line diagram for your figuring.

AREA TAKEOFF

The takeoff of the area is actually very easy once you know how to do it. Here are a few things to remember for doing this takeoff.

1. The best and easiest way to measure the area of the building is to break the building down into rectangles, squares, triangles, circles and semicircles. You can then calculate the areas of all these parts separately and add them up to get a total figure.

2. Always measure the outside dimensions of the building. There is nothing wrong with using the inside dimensions, but that makes the task many times harder and won't give you any real benefits.

3. Make sure you include all levels of the building. If the project you are working on is a multilevel building, don't forget the upper stories. (This is elementary, but it's worth repeating.)

4. Do not include areas that get no wiring, such as covered walkways, canopy areas, and so on.

5. Do not rely on a square footage measurement supplied by the architect. This is an important part of your takeoff, and it is not wise to trust someone else's measurements. Some of the architect's reported square footage measurements are very accurate, but many (maybe the majority) are not. The little bit of time that you will spend figuring the square footage yourself is time well spent.

Here are the basic formulas that you need to know to calculate the areas of different shapes:

Rectangles: Multiply the length of one side by the length of an adjacent side.

Squares: Same operation as for rectangles.

Triangles: The area of a triangle is equal to one-half the length of the base times the height.

Circles: To calculate the area of a circle, multiply the square of the radius (radius x radius) times "pi" (π). (The radius of a circle is the distance from the center of the circle to the edge. The number called "pi" equals 3.1416.)

Semicircles: Figure the area as if the semicircle were a full circle (radius squared times pi) and then divide by two.

SPECIAL SYSTEMS ESTIMATING

Special wiring systems such as fire alarm systems, security systems, sound systems, special TV systems, special grounding systems and lightning protection are categorized as special systems.

Electrical contractors typically deal with special systems in two different ways:

First, they may use specialty subcontractors to handle their special systems installations. This is how it usually works. The estimator does a review of the project that is out for bid. He notices that there are a couple of special systems that will be required — a fire alarm system and a sound system. He calls a few of the specialty contractors he has used before for these types of systems and asks them to figure the job for him (and verifies that they indeed will). At some time before the bid is due, these specialty contractors will either call their bids in to the estimator or mail them. The estimator will then take the best bids and figure them into the job.

When the job is awarded, the specialty contractor is given a purchase order and is expected to provide as many sets of submittals as are required. After the submittals are approved and the job is ready, the specialty contractor will come to the job site and install the system.

Some electrical contractors like to do their own special systems installations. They feel that they can save enough money on the installation to make it worth their while. Thus a second way that electrical contractors deal with special systems is described below.

After the estimator completes his regular takeoff, he

SPECIAL SYSTEMS ESTIMATING *(cont.)*

does a takeoff of the special system(s) and comes up with figures for total material cost and labor cost for the installation. (Often, but not always, this requires a couple of phone calls to the manufacturer.) Of course, the estimator then adds a certain percentage for overhead and profit and tabulates the bid.

After the job is awarded, the electrical contractor is responsible for doing all his own submittals and, of course, the installation itself. While contractors can save some money by doing the special systems work themselves, generally the disadvantages far outweigh the advantages.

FINAL REVIEW

In your final review, you must take into account any odd parts of the job and any particular job situation that you can foresee.

"Odd parts of the job" means any unusual work that will be required for this project and would not normally be done on this type of job.

For any unusual type of installation you need to come up with a total dollar value (material, labor, overhead and profit).

Some common additional expenses you might foresee are:

Storage trailers

Extra insurance on trailers and their contents

Utility costs (phone and/or electric)

Freight expenses

Lodging and food expenses

FINAL REVIEW *(cont.)*

Equipment rentals

Gas, oil and maintenance for equipment on the job

Interest on any loans that will be required to finance the project

Cost of a nonproductive foreman where required

Any other expenses required by this job, but not included in your company's overhead

As with the odd parts of the job, you need to come up with a dollar amount for any of these job expenses and plug them into your estimate.

Unless the special systems are very extensive, there is no need to do a takeoff of empty conduits that may be required for the special systems work. If you do run into a job with very extensive special systems work, make a takeoff of the empty conduit system and account for it as you would for any other unusual parts of the job.

After you finish these six steps, your takeoff is complete. There is nothing else for you to do as far as taking off the job is concerned. As we have said, the biggest savings in time that the algebraic estimating method provides is that it enables you to do only a partial takeoff while still retaining accuracy.

CALCULATIONS

With your takeoff complete, the rest of the estimate is done with calculations. The algebraic method uses about eight basic formulas for its operations.

CALCULATIONS *(cont.)*

We'll start with our primary form, the Calculation Sheet, which is shown in **Figure 5.1 (page 5-44)**. At the top you enter the name of the person who prepares the estimate and the date. Go on to enter the name of the project, the total area of the job and the bid date.

In the next section you fill in your raw cost data. Fixture dollars and gear dollars (gear is short for switchgear) will typically be bid to you by several suppliers on bid day. You take the lowest responsible bid and enter the numbers here. Fixture hours and gear hours are the figures that will come from your takeoff of light fixtures and the one-line diagram. Your figures for total labor hours needed for these installations will go here.

The next step is to calculate and fill in the figures for dollars per thousand square feet ($/MSF) and hours per thousand square feet (Hrs/MSF). To get these figures, all you have to do is to divide the total dollars or hours by how many thousand square feet there are in the job (total job area).

For example, fixture and switchgear costs for a certain project are as follows:

Fixture $: $41,450.00 Fixture Hours: 971

Gear $: $18,754.00 Gear Hours: 242

Total Job Area: 42,791 square feet

Therefore, Fixture $/MSF would be $41,450.00 divided by 42.791, which comes to $968.66/MSF.

Fixture Hrs/MSF would be 971 divided by 42.791, which comes to 22.69 Hrs/MSF.

CALCULATIONS *(cont.)*

Gear $/MSF would be $18,754.00 divided by 42.791, which is $438.27/MSF.

Gear Hrs/MSF would be 242 divided by 42.791, which equals 5.655 Hrs/MSF.

The next section of the Calculation Sheet is for filling in information from a comparison job. You should choose a good comparison job to use in relation to the job that is bidding. (This requires you to keep a file of previous jobs and estimates, as you should anyway.)

In this section of the Calculation Sheet, fill in the name of the comparison job and the date that it was bid. Next fill in these figures for the comparison job:

Fixture $/MSF

Fixture Hrs/MSF

Gear $/MSF

Gear Hrs/MSF

The total cost of the comparison job per square foot

Now you must do a comparison between the job being bid and the comparison job. The best formula for this (and for many of our other calculations) is:

$$\frac{\text{Job Being Bid}}{\text{Comparison Job}} = \text{Comparison}$$

Let's say that the comparison job has these figures:

Fixture $/MSF = $951.84

Fixture Hrs/MSF = 22.74

Gear $/MSF = $441.90

Gear Hrs/MSF = 5.721

Total cost per square foot = $2.974

CALCULATIONS *(cont.)*

Notice that for the total cost figure we use not only dollars and cents but even tenths of a cent ($2.97<u>4</u>). Remember that always using four significant digits helps maintain accuracy. This is why we go down to tenths of a cent.

Comparing these figures to the figures given earlier for the example job being bid, we can calculate these comparisons as follows:

Fixture $: $968.66/MSF
divided by $951.84/MSF = 1.0177

Fixture Hrs: 22.69 Hrs/MSF
divided by 22.74 Hrs/MSF = .9978

Gear $: $438.27/MSF
divided by $441.90/MSF = .9918

Gear Hrs: 5.655 Hrs/MSF
divided by 5.721 Hrs/MSF = .9885

The next step is to get a combined (or "combo") figure for both fixtures and gear. This is a very simple average of the two comparison figures. The formula is:

$$\frac{\$ \text{ comparison figure} + \text{hrs. comparison figure}}{2}$$

For this example, this is what they would be:

Fixtures: 1.0177 + .9978 = 2.0155
2.0155 divided by 2 = 1.0078

Gear: .9918 + .9885 = 1.9803
1.9803 divided by 2 = .9902

CALCULATIONS *(cont.)*

Note: Throughout the system's calculations, the numbers used are based upon their dollar values, but for this calculation, we use only a simple average rather than giving the materials more weight than labor, based upon material's greater dollar worth. This is done because these numbers are being used here as adjusting numbers, not as cost figures. The material values and labor values are of equal importance in adjusting the job, regardless of their dollar worth.

The next step in the process is figuring the Total Job Coefficient. The worksheet for figuring the Total Job Coefficient is shown in **Figure 5.2 (page 5-45)**.

The first step is to figure the total installed cost of the light fixtures and the switchgear. This is done by adding the material cost to the total labor cost. (Note that for this calculation use labor dollars, not the labor hours, which we have typically used.) To get the total labor cost, multiply the number of labor hours times the average hourly cost, an average hourly rate of $15.00 per hour. [The hourly rate should include all taxes and insurance and any other expense that is directly tied to labor. This would typically include items such as FICA (Social Security), unemployment taxes, workmen's compensation insurance, union funds, contractor's associations, apprenticeship training funds, etc.]

Taking the fixture cost of $41,450.00 and adding to it the total labor cost for installing the fixtures of $14,565.00, we get a total installed cost of $56,015.00.

The term "installed cost" is abbreviated "IC."

Now take the gear cost of $18,754.00 and add to it the total labor cost of $3,630.00 (242 hours x $15.00/hour) to get a total installed cost for the gear of $22,384.00.

Divide the Fixture IC (Fixture Installed Cost) by Gear IC (Gear Installed Cost). This gives a ratio of fixtures to gear. This is essential to getting a properly weighted multiplier. This is abbreviated "F/G Ratio."

For the next step, you have to go back to the Calculation Sheet **(Figure 5.1)** to get a couple of figures: Fixture Combo and Gear Combo.

Here's what to do:

1. Multiply the Fixture Combo times the F/G Ratio.
2. Add to this figure the Gear Combo.
3. Divide this total by the F/G Ratio plus 1.

You will note that this is the algebraic equation for a balanced multiplier that was introduced earlier in this chapter.

Here are the calculations for our example:.

1. Divide the Fixture IC by the Gear IC; that is, $56,015.00 divided by $22,384.00 = 2.5025. This figure (2.5025) is the F/G Ratio. Now take the Fixture Combo (1.0078) and multiply it by the F/G Ratio (2.5025) to yield 2.522.
2. To this amount (2.522) add the Gear Combo (.9902) to get 3.5122.

CALCULATIONS *(cont.)*

3. Divide this total by the F/G Ratio plus 1 to get the Total Job Coefficient:
   3.5122 divided by 3.5025 (F/G Ratio + 1) = 1.0028

The Total Job Coefficient is 1.0028.

Now let's return to the Calculation Sheet. The next step after obtaining the Total Job Coefficient is to multiply the Total Job Coefficient times the Total Price per Square Foot of the comparison job. This gives a Total Price per Square Foot for the job being bid.

Multiply this price per square foot times the number of square feet in the job being bid. This gives the bid price.

Again, let's do these calculations for our example:

Total Job Coefficient (1.0028) times the cost per square foot of the comparison job ($2.974) = total price per square foot for the job being bid (2.9823).

Total price per square foot for the job being bid (2.9823) times the number of square feet in the job being bid (42,791) = the bid price (127,615.60).

At this point, the estimate is complete if you don't need to make any adjustments. However, you will usually have to make one or two adjustments.

ADJUSTMENTS

With the basic estimate complete, the next step is to make any adjustments required. The adjustments are:

1. **Special systems adjustment.**
2. **Inflation adjustment.**
3. **Labor rate adjustment.**
4. **Job cost adjustment.**
5. **Other adjustments.**
6. **Overhead and profit adjustment.**

We will continue to use the same example for the adjustments as we did for the calculations.

Special Systems

As mentioned before, special systems vary greatly from one project to another, so a special adjustment process is used to properly account for the special systems.

First, review the comparison job and find the total cost per square foot for the special systems. If you used a Jobs Data Sheet like the one shown in **Figure 5.9 (page 5-52)**, you will find this information already calculated and recorded. If not, you will have to divide the price of the special systems (Material Cost + Labor Cost + Sales Tax + Overhead and Profit) by the number of square feet in the job.

Since you are using the comparison job's price per square foot and adjusting it to the job being bid, you need to know how much of this per square foot cost is special systems. (What you are doing is calculating the amount per square foot for special systems so you can pull the special systems cost out of the estimate completely.)

Multiply the price per square foot of special systems in the comparison job by the number of square feet in the job being bid. This shows how much you have already figured into the job being bid for special systems.

Last, take your quotes for special systems on the new job, find the difference between these quotes and the amount that you have already figured into the bid price, and add or subtract this amount from the bid price.

Figure 5.3 (page 5-46) shows a Special Systems Adjustment form. The top section is for calculating the amount of special systems already figured into the bid price. In the middle is a section for totaling all your special systems quotes (or estimates). At the bottom is a section for figuring the difference. To avoid any problems, check off whether it is an add or deduct.

Continuing the example, this is how a special systems adjustment works:

We begin with a special systems cost of $0.3714/square foot.

Multiply this figure by the number of square feet in the job being bid — 42,791 — to get the answer of $15,892.58; this means that we have already figured $15,892.58 into the bid price for special systems.

Now total all our quotes for special systems on the job being bid to come up with $10,971.00 (this includes any applicable sales tax).

Take the difference between these two numbers. This comes out to $4,921.58. Since there was more money figured into the job than the special systems will actually cost, this amount will have to be deducted from the bid amount. If this amount were greater, the difference would need to be added to the bid amount.

Inflation

Another adjustment is designed to account for inflation. To make this adjustment you must first calculate how much inflation there has been from the time the comparison job was either bid or calculated until the time at which you are bidding the job you are adjusting.

The best gauge of inflation is the "Finished Goods" portion of the "Producer Price Index," which is published by the U.S. Government Bureau of Labor Statistics. The easiest place to find this information is in the magazine *U.S. News & World Report*. It can be found in the section of the magazine titled "Monthly Indicators" (Producer Prices, Finished Goods). This is in the part of the magazine called USN & WR Index of Business Activity. This information can also be found at almost any public library.

To calculate the percentage of inflation, go back to the basic formula:

$$\frac{\text{Job Being Bid}}{\text{Comparison Job}} = \text{Multiplier}$$

Figure 5.4 (page 5-47) shows the Inflation and Labor Adjustment form. At the top you see the bid dates and Index level for both the job being bid and the comparison job. Notice that you divide the Index level of the job being bid by the Index level of the comparison job. This gives a Material Adjustment Multiplier, or inflation multiplier.

Continue with the example job: At the time our comparison job was bid, the Index was at 271.4 and at the time you are bidding this job, the Index is at 281.0. Divide 281.0 by 271.4 (job being bid divided by comparison job) to get a multiplier of 1.0354 (a 3.54 percent increase).

Labor

The labor adjustment follows the same pattern as the inflation adjustment, as you can see on the form in **Figure 5.4**. Divide the labor rate for the job you are bidding by the labor rate of the comparison job. This gives your Labor Adjustment Multiplier.

Continue the example: At the time the comparison job was bid, the hourly labor rate was $14.74; the current hourly labor rate is $15.00.

Divide the labor rate for the job being bid ($15.00) by the labor rate of the comparison job ($14.74). This gives a multiplier of 1.0176, or a labor cost 1.76 percent greater.

Integrating Labor and Materials

You now have the inflation multiplier and the labor adjustment multiplier. The inflation multiplier covers materials only, while the labor adjustment obviously covers labor only.

The next step is to integrate these two multipliers into one. To do this, go back to the algebraic equation that was used earlier.

Here is how the process goes:

First add the fixture dollars and gear dollars to get a material cost. Then add the cost of the fixture installation and the cost of the switchgear installation to get a labor cost.

Divide the material cost (in dollars) by the labor cost (also in dollars) to get a Material to Labor ratio (M/L ratio).

This brings us to the formula. For simplicity's sake, we will call the Inflation Multiplier "IM" and the Labor Multiplier "LM." This is what the formula looks like:

$$\frac{(\text{M/L ratio} \times \text{IM}) + \text{LM}}{\text{M/L ratio} + 1} = \text{Integrated Multiplier}$$

After obtaining this multiplier, you will add or subtract from the bid price to adjust it. Since 1.0 indicates virtually no change at all, take the difference between 1.0 and the Integrated Multiplier. Multiply this figure by the bid price. If the Integrated Multiplier is greater than 1.0, add this sum to the bid price. If the Integrated Multiplier is less than 1.0, deduct this sum from the bid price.

Looking at the Integrated Multiplier sheet shown in **Figure 5.5 (page 5-48)**, you see at the top the section for the calculations of Material Cost and Labor Cost. In the middle section you see the Material Cost divided by the Labor Cost to yield the M/L (Material to Labor) ratio. Then you see the application of the formula: multiply the Inflation Multiplier by the M/L ratio. To this sum add the Labor Multiplier, then divide this whole sum by the M/L ratio plus 1 to yield the Integrated Multiplier.

To arrive at the final dollar figure to add or deduct from the bid price, take the difference between the Integrated Multiplier and 1.0 and multiply this difference by the bid price.

Now let's continue the example through this step:

If you refer back to the Total Job Coefficient sheet in **Figure 5.2**, you will find material and installation costs for light fixtures and switchgear. You find that when you add the fixture cost of $41,450.00 and the switchgear cost of $18,754.00 you get a total material cost of $60,204.00.

Add the cost of the fixture installation ($14,565.00) and the cost of the switchgear installation ($3,630.00) to get a total labor cost of $18,195.00.

Divide the material cost by the labor cost to get the M/L ratio: $60,204.00 divided by $18,195.00 = 3.3088.

Apply the formula

$$\frac{(\text{M/L ratio} \times \text{IM}) + \text{LM}}{\text{M/L ratio} + 1} = \text{Integrated Multiplier}$$

by filling in the known values:

$$\frac{(3.3088 \times 1.0354) + 1.0176}{3.3088 + 1} = \text{Integrated Multiplier}$$

Complete the operations:

$$\frac{3.426 + 1.0176}{4.3088} = \text{Integrated Multiplier}$$

$$\frac{4.444}{4.3088} = 1.0314$$

The Integrated Multiplier is 1.0314, or 103.14 percent. Since the bid price that we already have equals 100 percent of itself, we will want to raise this price by 3.14 percent.

.0314 (3.14 percent) x 127,615.60 = 4.007.13, so we need to adjust the bid price upward by the amount of 4,007.13.

It is important to note that if only one of the two parts (material due to inflation or labor due to rate changes) is being adjusted, you should do all the calculations just as you would if they were both being adjusted. The only difference is that instead of using the multiplier of the section *not* being adjusted, you will simply use 1.

Job Cost

The next adjustment is the job cost adjustment, which is virtually the same as the special systems adjustment.

First, calculate the cost per square foot of the comparison job and be sure to include any sales, overhead and profit. (If you used a Jobs Data Sheet like the one in **Figure 5.9**, you will find this information already recorded. We'll come to this shortly.)

Multiply this per square foot cost by the number of square feet in the job being bid. This shows how much has already been figured into the bid price.

Calculate the job cost of the job being bid. This figure should also include any sales tax, overhead and profit. Make sure, however, that when you add overhead and profit you use the overhead and profit percentages from the comparison job rather than those from the job being bid. This is because the entire job's overhead and profit will be covered by the overhead and profit adjustment.

Last, take the difference between these two numbers and add or subtract the difference from the bid price.

Figure 5.6 (page 5-49) shows a sheet for working out the Job Cost Adjustment. You can see how similar it is to the Special Systems Adjustment sheet.

Now let's continue with our example:

First, determine that the total job cost for the comparison job came to $0.059 per square foot (this includes overhead, profit, etc.).

Multiply this by the number of square feet in the job being bid to ascertain the total amount for job cost that you have already figured into the bid price. $0.059 per square foot x 42,791 square feet = $2,524.67.

Add up all the job costs for the job being bid and add overhead, profit, and so on (remember to use the percentage of overhead and profit from the comparison job) to get the total job costs for the job being bid. This comes out to $3,024.50.

Last, take the difference between these two numbers ($3,024.50 - $2,524.67 = $499.83) and add this figure to your bid price, for the final job cost of the job being bid is greater than the job cost based upon the comparison job.

If the job cost taken from the comparison job had been greater than that of the job being bid, you would deduct this amount from the bid price rather than adding it.

Other Adjustments

If during the course of an estimate you find any other peculiar parts of a job that must be accounted for, you will have to add them into the job or pull them out. In any event, the formulas shown here will be all that you need.

The only time that this can get tricky at all is when you have a certain peculiar part of one of your comparison jobs that you will need to pull out. You do this just the same as you would for the special systems adjustment or the job cost adjustment. You figure the amount that this peculiar part of the job cost per square foot (including overhead and profit), multiply the square foot cost by the square footage of the job being bid, and then deduct this amount from the bid price.

Overhead and Profit

This brings us to the last adjustment: the overhead and profit adjustment. The overhead and profit adjustment uses the basic formula:

$$\frac{\text{Job Being Bid}}{\text{Comparison Job}} = \text{Multiplier}$$

ADJUSTMENTS *(cont.)*

First, find the percentage of overhead and profit used for the comparison job.

Next, decide on what percentage of overhead and profit you want to use for the job being bid.

Divide 1 plus the percentage you are using for the job being bid by 1 plus the percentage that was used in the comparison job. This gives you an accurate ratio of one job to the other.

Take the difference between this number and 1.0 (which is the percentage of increase or decrease) and multiply it by an adjusted bid price to find out how much this adjusted bid price should be raised or lowered.

You total the bid price with all other adjustments and adjust the overhead and profit for all of them at the same time. This is why the overhead and profit adjustment is done last.

The form for doing your Overhead and Profit Adjustment, as shown in **Figure 5.7 (page 5-50)**, is fairly simple.

Let's finish the example:

The overhead and profit percentage used in the comparison job was 21 percent. You decide that for the job that is now bidding, you want to use a little higher percentage of overhead and profit — 24 percent.

Divide 1 plus the percentage you are using for the job being bid (1.24) by 1 plus the percentage that was used in the comparison job (1.21). This gives a multiplier of 1.0248.

ADJUSTMENTS *(cont.)*

Find the difference between this number and 1.0, which is .0248, or an increase of 2.48 percent.

Total your original bid price (127,615.60) with all the other adjustments that you have made to get an adjusted bid price:

Bid price	$127,615.60
Special systems adjustment	[4,921.58]
Inflation and labor adjustment	4,007.13
Job cost adjustment	499.83
Adjusted bid price	$127,200.98

Notice that numbers shown in brackets [ ] are deducted.

Multiply .0248 by the adjusted bid price (127,200.98) which also can be labeled the subtotal, to get the overhead and profit adjustment of $3,154.58.

Figure 5.8 (page 5-51) shows a Bid Sheet, where these final steps are completed.

At this point your estimate is complete. In actual practice it would be unusual to use all these adjustments on one project, so your estimating process will actually be easier than what is shown here.

JOBS DATA FILE

Because of its design, algebraic estimating is dependent upon the use of comparison jobs. This requires an electrical contractor to keep a jobs data file. This is simply a record of all significant projects that have been done or bid in the past. All electrical contractors should keep such a file whether they use algebraic estimating or not. They often do not because of the "hurry-up" nature of their jobs. Algebraic estimating *requires* you to maintain your file rather than giving you an option.

Figure 5.9 (page 5-52) shows a copy of a Jobs Data Sheet. This sheet contains the typical information that is worthwhile for an electrical contractor to save. In addition, the sheet contains information that is tailored to algebraic-type estimates.

At the top of the sheet is a place for the preparer's name and the date that the sheet was prepared. Below this is a place for the project name and the location. The location of the job has no direct significance to the estimate, but it is included as a safety check because productivity, labor rates, material prices, job costs and so on, will vary considerably from one area to another.

Next are places for the bid date, bid price, inflation index, cost per square foot, square footage, labor rate, overhead and profit percentage and job costs.

The inflation index is the Index of Leading Economic Indicators that was discussed earlier. The overhead and profit percentage is the amount the job has been marked up. To calculate this, you divide the selling price by the total cost.

In the description section you should record any unusual parts of the job. Remember also to include the cost of these unusual types of items so that they can be properly accounted for when you figure new projects.

JOBS DATA FILE *(cont.)*

If you keep a good jobs data file and categorize the entries by type of project (schools, hospitals, offices, supermarkets, retail stores, etc.), it can be of great value to you. Once you have a good jobs data file and learn how to use it well, you will probably find yourself using it almost daily.

As with all your figures, try to always use at least four significant digits for every entry on the jobs data sheet. This keeps your percentage of error at a minimum (a maximum of 0.0049 percent). Once you have an estimate complete, the Jobs Data Sheet should only take a few minutes to fill out. Just be sure to take an extra minute to review the project, and be sure there are no significant parts of the job that are not well noted.

NOTES

It is very fortunate that this method of estimating works so well for electrical construction projects. Had any one of several factors been different, the resulting system would be somewhat less effective. This technique would still be useful, but the whole system would not be able to work together as it does now.

There are several differences between the algebraic estimate and the conventional itemized estimate that you should remember.

First, in the algebraic type of estimate, the key to accuracy is correct ratios and comparisons: the concept of intersectional proportionality. In the itemized method, the key to accuracy is correctly counting and tabulating all the individual items.

NOTES *(cont.)*

Because of this, the algebraic type of estimating gives you no room to "fudge" numbers. Almost all electrical estimators are familiar with "guesstimating" certain parts of the job when under pressure. With algebraic estimating you have a lot less leeway to do so. If you were to guesstimate a ratio or multiplier and guess wrong, you could get yourself in a lot more trouble than you would by guesstimating the wrong number of wire nuts.

It is also important to remember that the algebraic type of estimate cannot be used all the time. The system is not applicable for utility work, partial renovations, or other odd types of jobs.

Depending on your location, an algebraic type of estimate can be used on between 75 percent and 95 percent of projects that go out for bid. Although it would be nice to use the system on 100 percent of the jobs, you just can't, as the system does have some limitations. In any case, the system can be used on the majority of jobs that come out for bid.

One very interesting benefit to using the algebraic system is that it gives you tremendous insight into the inner workings of an electrical project. After an estimator becomes familiar with the system and has used it a number of times, he or she gets a much better feel for the anatomy of a job — its inner workings and how one part of a job relates to another. It is this kind of insight that makes the difference between a capable estimator and a seasoned veteran who can "feel" the job, not just count the parts.

NOTES *(cont.)*

All the dollar cost figures used in the system should include sales tax. If not, there is a certain amount of "slop" in some of the calculations, particularly in some of the adjustments. For all your figures and jobs data, include sales tax in the dollar cost. *Note:* If there is a sales tax increase, handle it like an inflation increase.

FIGURE 5.1 — CALCULATION SHEET

Calculation Sheet

Prepared By: ______________________ Date: ____________

Job Name: ____

Total Job Area: _______________ Bid Date: ____________

Fixtures: $ _________________ Hours: _______________

Gear: $ ____________________ Hours: _______________

Fixtures — $/MSF: $ ________ Hrs/MSF: ____________

Gear — $/MSF: $ ________ Hrs/MSF: ____________

Comparison Job: _____________ Date Bid: ____________

Fixtures — $/MSF: $ ________ Hrs/MSF: ____________

Gear — $/MSF: $ ________ Hrs/MSF: ____________

Price Per Sq. Ft.: $ __________

Comparison: Job Being Bid Divided by Comparison Job

Fixtures: $ _________ Hrs: __________ Combo: ________

Gear: $ _________ Hrs: __________ Combo: ________

Total Job Coefficient: __________

Total Job Coeff. x Total Price/Sq. Ft. (Comparison)

= Price/Sq. Ft.

__________ x ___________ = $ ___________

Price/Sq. Ft. x Sq. Ft. of Job Being Bid = Bid Price

__________ x ___________ = $ ___________

FIGURE 5.2 — TOTAL JOB COEFFICIENT

Total Job Coefficient

Fixtures:

Fixture Cost . . . . $ ______________

Hours @ $ ________ /Hr . . . . . . $ ______________

Fixture IC $ ______________

Gear:

Gear Cost . . . . $ ______________

Hours @ $ ________ /Hr . . . . . . $ ______________

Gear IC $ ______________

Fixture IC ________ /Gear IC________ = F/G Ratio______

From Calculation Sheet:

F/G Ratio ________ x Fixt. Combo________ = ________

Add Gear Combo: + ________

= ______(A)

(A)______________ Divided By F/G Ratio + 1 ________

= Total Job Coefficient ______________

FIGURE 5.3 — SPECIAL SYSTEMS ADJUSTMENT

Special Systems Adjustment

Special Systems Cost/Sq. Ft. in Comparison Job: ______

Number of Square Feet in Job Being Bid: ____________

Cost/Sq. Ft.: _____ x Number of Sq. Ft.: _____ = _____ (A)

Special Systems Quotes for Job Being Bid

	System	Vendor/ Salesman	Price
1.	____________	______________	$____________
2.	____________	______________	$____________
3.	____________	______________	$____________
4.	____________	______________	$____________
5.	____________	______________	$____________
		Total	$____________

Amount Figured into Bid Price: $ ________________ (A)

Total Special Systems Quotes: $ ________________ (B)

Difference between (A) & (B): $ ________________

❑ Add to Bid

❑ Deduct from Bid

FIGURE 5.4 — INFLATION AND LABOR ADJUSTMENT

Inflation and Labor Adjustment

Material

Job Being Bid		Comparison Job	
Date: ________________		Date: ________________	
Index: ________________	(A)	Index: ________________	(B)

(A) ________________ Divided by (B) ________________

= Material Adjustment Multiplier __________

Labor

Job Being Bid		Comparison Job	
Date:. ________________		Date: ________________	
Labor rate: ______________	(A)	Labor rate ______________	(B)

(A) ________________ Divided by (B) ________________

= Labor Adjustment Multiplier ____________

FIGURE 5.5 — INTEGRATED MULTIPLIER

Integrated Multiplier

Fixtures . . . . $__________

Gear . . . . $__________

Material Cost: $__________

Fixture Hours @ $ _______ /Hr . . . . $__________

Gear Hours @ $ _________ /Hr . . . . $__________

Labor: $__________

Material Cost $ ______ Divided by Labor Cost $ ______

= M/L Ratio: ____________

M/L Ratio __________ x IM ___________ = __________

+ LM ______

= __________(A)

(A) ___________ Divided by M/L Ratio + 1 __________

= Integrated Multiplier __________

Take the difference between the Integrated Multiplier and 1.0 and enter it at (B).

(B) ________ x Bid Price ________ = Adjustment ________

❑ Add to Bid ❑ Deduct from Bid

FIGURE 5.6 — JOB COST ADJUSTMENT

Job Cost Adjustment

Job Cost/Sq. Ft. in Comparison Job:________________

Number of Square Feet in Job Being Bid: ____________

Cost/Sq. Ft. _____ x Number of Sq. Ft. _____= _____ (A)

Job Costs for Job Being Bid

	Item	Vendor	Price
1.	____________	______________	$____________
2.	____________	______________	$____________
3.	____________	______________	$____________
4.	____________	______________	$____________
5.	____________	______________	$____________
6.	____________	______________	$____________
7.	____________	______________	$____________
		Total	$____________

Amount Figured into Bid Price: $ ________________ (A)

Total Job Cost: $ ________________ (B)

Difference between (A) & (B): $ ________________

❑ Add to Bid ❑ Deduct from Bid

FIGURE 5.7 — OVERHEAD AND PROFIT ADJUSTMENT

Overhead and Profit Adjustment

O & P Percentage for Job Being Bid_______________ (A)

O & P Percentage for Comparison Job_____________ (B)

(A) _______________Divided by (B) _______________

= O & P Multiplier ____________________

Price to be Adjusted by _________________%

Adjusted Price _______ x Adjustment Multiplier ______

= O & P Adjustment __________________

❑ Raise Price ❑ Lower Price

FIGURE 5.8 — BID SHEET

Bid Sheet

Bid Price	$__________
Special Systems Adjustment	$__________
Inflation and Labor Adjustment	$__________
Job Cost Adjustment	$__________
Other: ______________________________	$__________
Other: ______________________________	$__________
Subtotal	$__________
Overhead and Profit Adjustment	$__________
Final Bid Price	$__________

FIGURE 5.9 — JOBS DATA SHEET

Jobs Data Sheet

Prepared By: ______________________ Date: ____________

Job Name: ____

Location: ____

Date Bid: ________________ Bid Price: $ _____________

Inflation Index: ___________ Cost/Sq. Ft. $ ___________

Square Footage: _________ Labor Rate: $ _____ /Hour

Ovhd. & Profit %: ________ Job Costs: $ ____________

Job Description: _______________________________________

____

____

____

____

Fixture $: _______________ $/MSF: _________________

Fixture Hrs: _____________ Hrs/MSF: _______________

Gear $: _________________ $/MSF: _________________

Gear Hrs: _______________ Hrs/MSF: _______________

Misc. Pipe & Wire $: _____ $/MSF: _________________

Misc. Pipe & Wire Hrs: ___ Hrs/MSF: _______________

Special Systems $: ______ Price/Sq. Ft.: ____________

Description of Systems:

______________: $ ______________ Hrs: _____________

______________: $ ______________ Hrs: _____________

______________: $ ______________ Hrs: _____________

______________: $ ______________ Hrs: _____________

NOTES

About The Author

Paul Rosenberg has an extensive background in the construction, data, electrical, HVAC and plumbing trades. He is a leading voice in the electrical industry with years of experience from an apprentice to a project manager. Paul has written for all of the leading electrical and low voltage industry magazines and has authored more than 30 books.

In addition, he wrote the first standard for the installation of optical cables (ANSI-NEIS-301) and was awarded a patent for a power transmission module. Paul currently serves as contributing editor for *Power Outlet Magazine*, teaches for Iowa State University and works as a consultant and expert witness in legal cases. He speaks occasionally at industry events.